Performance and Constraint Satisfaction in Robust Economic Model Predictive Control

Von der Fakultät Konstruktions-, Produktions- und Fahrzeugtechnik
und dem Stuttgart Research Centre for Simulation Technology
der Universität Stuttgart zur Erlangung der Würde eines
Doktor-Ingenieurs (Dr.-Ing.) genehmigte Abhandlung

Vorgelegt von

Florian Anton Bayer

aus Ostfildern

Hauptberichter: Prof. Dr.-Ing. Frank Allgöwer
Mitberichter: Prof. Daniel Limón Marruedo, Ph.D.
Prof. Daniel E. Quevedo, Ph.D.

Tag der mündlichen Prüfung: 18. September 2017

Institut für Systemtheorie und Regelungstechnik
Universität Stuttgart
2017

Bibliografische Information der Deutschen Nationalbibliothek

Die Deutsche Nationalbibliothek verzeichnet diese Publikation in der
Deutschen Nationalbibliografie; detaillierte bibliografische Daten sind
im Internet über http://dnb.d-nb.de abrufbar.

D93

ISBN 978-3-8325-4573-4

Logos Verlag Berlin GmbH
Comeniushof, Gubener Str. 47,
10243 Berlin
Tel.: +49 (0)30 42 85 10 90
Fax: +49 (0)30 42 85 10 92
INTERNET: http://www.logos-verlag.de

Für meine Familie

Acknowledgements

This thesis presents the results which I obtained during my time as research and teaching assistant at the Institute for Systems Theory and Automatic Control (IST) of the University of Stuttgart. There are many people who helped me during this time and who were involved in my research.

First of all, I would like to express my gratitude to my advisor Prof. Frank Allgöwer for his support, mentoring, and his valuable advice during my time at the IST. He created a fruitful and unique research environment and gave me the freedom to pursue my own research ideas. Moreover, he provided the opportunity to meet and discuss with researchers in my field from all over the world and to attend various conferences to become involved in the scientific community.

I would also like to thank Prof. Daniel Limón Marruedo (University of Seville), Prof. Daniel Quevedo (University of Paderborn), and Prof. Peter Eberhard (University of Stuttgart) for the interest in my work, for their valuable comments, and for taking part in my doctoral examination committee.

I am indebted to Prof. John Hauser (University of Colorado) for inviting me to Boulder when I still was an undergraduate student and for leading me the way to the research world. I also thank Prof. Giuseppe Notarstefano (University of Salento), who I met on this way, for distracting and challenging me with interesting questions from the field of optimal control.

Prof. James Rawlings (University of Wisconsin) invited me to spend three months in his group in Madison. I am thankful for the interesting discussions and for his support.

I would like to thank all of my colleagues at the IST for creating a unique and inspiring atmosphere. Special thanks go to my long-time office mate Christian Feller for all the discussions and support. Moreover, I would like to thank Florian Brunner, Mathias Bürger, Hans-Bernd Dürr, Gregor Goebel, Philipp Köhler, Matthias Lorenzen, Simon Michalowsky, Max Montenbruck, Matthias Müller, Georg Seyboth, and all the current and former members of the MPC group for the fruitful and valuable discussions and for the excellent collaboration. I thank Isabel Bayer, Christian Feller, Gregor Goebel, Sven Knüfer, Johannes Köhler, Philipp Köhler, Matthias Lorenzen, and Matthias Müller for proof-reading the first version of this thesis.

Last but not least I would like to thank my parents for all their unlimited love and support during my whole life. Thanks to all my family and friends, to my sister, and my loving girlfriend. All of them contributed to this thesis and helped me to achieve this goal.

Neuhausen auf den Fildern, October 2016

Florian A. Bayer

Table of Contents

Notation

The following list provides an overview of the notation used throughout the thesis. Additional symbols that are only used in certain sections are defined locally.

$\mathbb{R}$	set of real numbers
$\mathbb{R}_{\geq 0}$	set of non-negative real numbers
$\mathbb{R}_{\leq 0}$	set of non-positive real numbers
$\mathbb{R}^n$	set of real-valued vectors with dimension n
$\mathbb{R}^n_{\leq 0}$	set of real-valued vectors with dimension n and negative components, i.e., $\mathbb{R}^n_{\leq 0} := \{y \in \mathbb{R}^n : y_i \leq 0, \text{ for all } i \in \mathbb{I}_{[1,n]}\}$
$\mathbb{R}^{n \times m}$	set of real-valued matrices with dimension $n \times m$
$[a, b]$	interval $\{x \in \mathbb{R} : a \leq x \leq b\}$, with $a, b \in \mathbb{R}$ fix
$\mathbb{I}$	set of integer numbers
$\mathbb{I}_{\geq a}$	set of all integer numbers greater than or equal to $a \in \mathbb{R}$
$\mathbb{I}_{> a}$	set of all integer numbers greater than $a \in \mathbb{R}$
$\mathbb{I}_{[a,b]}$	set of all integer numbers on the interval $[a, b] \subseteq \mathbb{R}$
$\leq, <, >, \geq$	for vectors $a \in \mathbb{R}^n$, inequalities are understood componentwise, i.e., $a > 0$ means $a_i > 0$ for all $i \in \mathbb{I}_{[1,n]}$ with $a = [a_1, \ldots, a_n]^\top$
I_n	identity matrix of size $n \times n$
$\mathcal{K}$	a continuous function $\alpha : \mathbb{R}_{\geq 0} \to \mathbb{R}_{\geq 0}$ is a class $\mathcal{K}$ function ($\alpha \in \mathcal{K}$) if it is strictly increasing and $\alpha(0) = 0$
$\mathcal{K}_\infty$	a continuous function $\alpha : \mathbb{R}_{\geq 0} \to \mathbb{R}_{\geq 0}$ is a class $\mathcal{K}_\infty$ function ($\alpha \in \mathcal{K}_\infty$) if $\alpha \in \mathcal{K}$, and furthermore, $\alpha(s) \to \infty$ as $s \to \infty$
$\mathcal{KL}$	a continuous function $\beta : \mathbb{R}_{\geq 0} \times \mathbb{I}_{\geq 0} \to \mathbb{R}_{\geq 0}$ is a class $\mathcal{KL}$ function ($\beta \in \mathcal{KL}$) if $\beta(\cdot, t)$ is a class $\mathcal{K}$ function for each fixed $t \in \mathbb{I}_{\geq 0}$, and $\beta(s, \cdot)$ is strictly decreasing with $\lim_{t \to \infty} \beta(s, t) = 0$ for each fixed $s \in \mathbb{R}_{\geq 0}$
$:=$	definition, i.e., $a := b$ means a is defined to be equal to b
$|x|$	Euclidean norm of $x \in \mathbb{R}^n$
$|x|_{\mathcal{A}}$	Euclidean set distance of $x \in \mathbb{R}^n$ to the set $\mathcal{A} \subseteq \mathbb{R}^n$, i.e., $|x|_{\mathcal{A}} := \inf_{a \in \mathcal{A}} |x - a|$
$\mathcal{B}_\epsilon(x)$	open ball of radius ϵ centered at $x \in \mathbb{R}^n$, i.e., $\mathcal{B}_\epsilon(x) := \{y \in \mathbb{R}^n : |x - y| < \epsilon\}$, $\mathcal{B}_\epsilon := \mathcal{B}_\epsilon(0)$
$\overline{\mathcal{B}}_\epsilon(x)$	closed ball of radius ϵ centered at $x \in \mathbb{R}^n$, i.e., $\overline{\mathcal{B}}_\epsilon(x) := \{y \in \mathbb{R}^n : |x - y| \leq \epsilon\}$, $\overline{\mathcal{B}}_\epsilon := \overline{\mathcal{B}}_\epsilon(0)$
$\text{int}(\mathcal{A})$	interior of a set $\mathcal{A} \subset \mathbb{R}^n$

AB multiplication of matrix $A \in \mathbb{R}^{n \times m}$ and set $\mathcal{B} \subseteq \mathbb{R}^m$, i.e.,
$$AB := \{Ab \in \mathbb{R}^n : b \in \mathcal{B}\}$$

$\mathcal{A} \oplus \mathcal{B}$ Minkowski set addition of $\mathcal{A} \subseteq \mathbb{R}^n$ and $\mathcal{B} \subseteq \mathbb{R}^n$, i.e.,
$$\mathcal{A} \oplus \mathcal{B} := \{a + b \in \mathbb{R}^n : a \in \mathcal{A}, b \in \mathcal{B}\}$$

$\mathcal{A} \ominus \mathcal{B}$ Pontryagin set difference of $\mathcal{A} \subseteq \mathbb{R}^n$ and $\mathcal{B} \subseteq \mathbb{R}^n$, i.e.,
$$\mathcal{A} \ominus \mathcal{B} := \{x \in \mathbb{R}^n : x + b \in \mathcal{A}, \forall b \in \mathcal{B}\}$$

$\lambda_{\max}(A)$ maximum eigenvalue of a symmetric matrix $A \in \mathbb{R}^{n \times n}$

$A \succ 0 \ (A \succeq 0)$ a matrix $A \in \mathbb{R}^{n \times n}$ is positive definite (positive semi-definite), i.e., A is symmetric and $x^\top A x > 0$ ($x^\top A x \geq 0$) for all $x \in \mathbb{R}^n$ with $x \neq 0$

$A \prec 0 \ (A \preceq 0)$ a matrix $A \in \mathbb{R}^{n \times n}$ is negative definite (negative semi-definite), i.e., A is symmetric and $x^\top A x < 0$ ($x^\top A x \leq 0$) for all $x \in \mathbb{R}^n$ with $x \neq 0$

$\mathcal{C}^k$ set of k-times continuously differentiable functions with $k \in \mathbb{I}_{\geq 0}$

$\boldsymbol{v}$ boldface letter represent finite sequences $\boldsymbol{v} : \mathbb{I}_{[0,N]} \to \mathbb{R}^n$ for some $N \in \mathbb{I}_{\geq 0}$, i.e, $\boldsymbol{v} = \{v(0), \ldots, v(N)\}$, or infinite sequences $\boldsymbol{v} : \mathbb{I}_{\geq 0} \to \mathbb{R}^n$, i.e., $\boldsymbol{v} = \{v(0), v(1), \ldots\}$, respectively

$\mathrm{Av}[\boldsymbol{v}]$ set of asymptotic averages of a bounded sequence $\boldsymbol{v} : \mathbb{I}_{\geq 0} \to \mathbb{R}^{n_v}$ defined by

$$\mathrm{Av}[\boldsymbol{v}] := \left\{ \bar{v} \in \mathbb{R}^{n_v} : \exists t_n \to \infty : \lim_{n \to \infty} \frac{1}{t_n + 1} \sum_{k=0}^{t_n - 1} v(k) = \bar{v} \right\}$$

where $\{t_n\}$ is an infinite subsequence of $\mathbb{I}_{\geq 0}$

Abstract

In this thesis, we develop a novel framework for model predictive control (MPC) which combines the concepts of robust MPC and economic MPC. *Robust MPC* takes disturbances into account to suppress their influence on the control task while satisfying imposed hard constraints. *Economic MPC* addresses arbitrary performance criteria, which can express the economic objectives of the underlying process to be controlled. The goal of this thesis is to develop and analyze MPC schemes for nonlinear discrete-time systems which explicitly consider the influence of disturbances on arbitrary performance criteria. Instead of regarding the two aspects separately, we propose *robust economic MPC* approaches that integrate information which is available about the disturbance directly into the economic framework. This may provide additional benefits, since tracking the nominally best behavior, i.e., the best behavior of the undisturbed system, need not provide the best closed-loop performance under economic criteria when the system is affected by disturbances.

In more detail, we develop three concepts which differ in which information about the disturbance is used and how this information is taken into account. While the first concept incorporates the disturbances in a worst-case manner, the second approach averages over all possible disturbances. The third concept takes additional stochastic information about the disturbance into account. Furthermore, we provide a thorough theoretical analysis for each of the three approaches. To this end, we present results on the asymptotic average performance as well as on optimal operating regimes. As is known from the disturbance-free case, optimal operating regimes are closely related to the notion of dissipativity, which is therefore analyzed for the presented concepts. Under suitable assumptions, results on necessity and sufficiency of dissipativity for optimal steady-state operation are established for all three robust economic MPC concepts. In addition, we present constructive approaches to choose the involved design parameters. A detailed discussion is provided which compares the different performance statements derived for the approaches, on the one hand, and the respective notions of dissipativity, on the other hand. The provided results are illustrated by numerical examples.

Deutsche Kurzfassung

Die modellprädiktive Regelung (englisch: *model predictive control*, MPC) ist ein modernes Regelungskonzept, das auf der wiederholten Lösung eines Optimalsteuerungsproblems mit endlichem Horizont basiert. Auf Grund seiner Flexibilität hinsichtlich der zu regelnden Systemklassen (Mehrgrößen- und nichtlineare Systeme sind möglich), der Möglichkeit Beschränkungen explizit zu berücksichtigen sowie der Freiheit beim Entwurf des zugrundeliegenden Gütekriteriums des Optimalsteuerungsproblems, hat MPC sowohl in der theoretischen Forschung als auch in der industriellen Praxis Beachtung gefunden.

Die vorliegende Arbeit entwickelt einen neuartigen MPC Ansatz, der die Ideen von robustem MPC mit ökonomischem MPC verbindet. Unter *robustem MPC* versteht man MPC Ansätze, die sowohl Stabilitätseigenschaften des geschlossenen Kreises als auch die Einhaltung gegebener Randbedigungen unter dem Einfluss von Störungen auf das System garantieren. Dazu berücksichtigen sie diese Störungen explizit im Entwurf des Optimalsteuerungsproblems. Das *ökonomische MPC* ist ein Verfahren, welches, im Gegensatz zum klassischen MPC, nicht auf die Stabilisierung eines gegebenen Sollwerts abzielt. Stattdessen liegt der Fokus auf der Minimierung eines allgemeinen Gütekriteriums, beispielsweise auf der Minimierung von Energiekosten oder der Profitmaximierung. Ziel der vorliegenden Arbeit ist die Entwicklung und die Analyse neuer MPC Verfahren für nichtlineare, zeitdiskrete Systeme, die explizit den Einfluss von Störungen auf allgemeine Gütekriterien berücksichtigen. Anstatt die beiden genannten Aspekte getrennt zu betrachten, werden *robuste ökonomische MPC* Verfahren vorgestellt, die verfügbare Informationen über die Störungen in ökonomisches MPC einfließen lassen. Dies kann zusätzliche Vorteile generieren, da sich die beste Regelgüte für das gestörte System nicht notwendigerweise einstellt, wenn man der optimalen Lösung des ungestörten Systems folgt.

Im Folgenden werden drei Konzepte entwickelt, die sich darin unterscheiden, welche Information über die Störung beim Reglerentwurf betrachtet wird und wie diese Information Verwendung findet. Das erste Konzept basiert auf min-max robustem MPC. Das heißt, es berücksichtigt die schlimmste mögliche Störung. Das zweite Konzept mittelt den Einfluss über alle möglichen Störungen. Das dritte Konzept berücksichtigt zusätzlich stochastische Information über die Störung. Jedes der drei Konzepte wird eingehend untersucht. Dabei werden Ergebnisse über die durchschnittliche Regelgüte sowie über das optimale Betriebsverhalten vorgestellt. Das optimale Betriebsverhalten ist, im ungestörten Fall, eng mit dem Prinzip der Dissipativität verknüpft, sodass Dissipativität unter Störungen untersucht wird. Für die drei robusten ökonomischen Konzepte wird gezeigt, dass Dissipativität sowohl notwendig als auch, unter gewissen Annahmen, hinreichend für den optimalen Betrieb des Systems an einer Gleichgewichtslage ist. Zudem werden konstruktive Ansätze für die Wahl von notwendigen Entwurfsparametern präsentiert. Die vorgestellten Verfahren werden sowohl bezüglich der unterschiedlichen Regelgüten, als auch im Hinblick auf ihre Dissipativitätsaussagen eingehend diskutiert und die gefundenen Ergebnisse mittels numerischer Simulationen untermauert.

Chapter 1

Introduction

1.1 Model predictive control

Many goals in control can be formulated in terms of (the solution of) an optimal control problem with infinite horizon. The resulting controller provides an input to achieve the desired task, but the infinite horizon optimal control problem can only be solved for special cases.

A modern approach to circumvent this drawback is Model Predictive Control (MPC). MPC is an optimization-based control technique which aims for approximating the solution of the infinite horizon optimal control problem by repeatedly solving an appropriately designed optimal control problem with *finite horizon*. The basic principle is as follows: At each time instance the current state is measured and the associated optimal control problem over a finite horizon is solved. The computed (optimal) predicted input is applied to the system for one time instance, see Figure 1.1. Then, the system state is re-evaluated and the optimization is started again. By considering the re-evaluated state of the system within the optimal control problem at each iteration, feedback is introduced.

MPC has successfully been applied to many problems as it (i) allows to consider a performance criterion within the underlying optimal control problem, (ii) can handle hard state and input constraints, and (iii) is applicable to nonlinear and multiple-input multiple-output systems. Due to these advantages, MPC has been of interest for researchers both from a theoretic and from an application-oriented point of view, and hence, the literature on theory as well as on successful applications is vast. Most of the approaches considered in the literature focus on stabilization (or tracking) of a desired behavior of the system to be controlled and are therefore referred to as "stabilizing MPC". The main theoretical aspects in stabilizing MPC are closed-loop stability as well as robustness against disturbances and

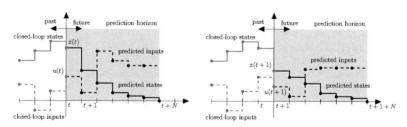

Figure 1.1: Basic concept of MPC.

uncertainties (see, e.g., Magni et al. (2009); Mayne et al. (2000); Rawlings and Mayne (2009), and the references therein).

Robust MPC

Considering robustness against disturbances and uncertainties has been of interest in the field of *stabilizing* MPC since the late 1980s (see, e.g., Campo and Morari (1987); Genceli and Nikolaou (1993); Michalska and Mayne (1993); Zheng and Morari (1993)), because most systems are either directly affected by disturbances or their representing model is subject to uncertainties. Two main directions become apparent in robust stabilizing MPC, (i) investigating inherent robustness and (ii) considering disturbances and uncertainties when designing the controller.

Inherent robustness is of particular interest when applying nominal MPC to real-world systems. Usually the system cannot be modeled exactly and/or disturbances act on the system. Hence, one would like to investigate if the underlying MPC algorithm designed for the undisturbed model provides stability and feasibility (see, e.g., Allan et al. (2016); Grimm et al. (2004, 2007); Pannocchia et al. (2011a,b); Teel (2004); Yu et al. (2011, 2014)). However, some information about the uncertainty and the disturbances is often available or can be approximated. Different approaches have been presented to explicitly incorporate knowledge about uncertainties and disturbances into the controller design, i.e., when designing the MPC algorithm and especially the underlying optimal control problem. The literature on these approaches is vast such that we only refer to some overview articles and books providing a starting point for further investigations (Bemporad and Morari, 1999; Grüne and Pannek, 2011; Kouvaritakis and Cannon, 2015; Mayne, 2014, 2016; Mayne et al., 2000; Rawlings and Mayne, 2009). The literature relevant for the results established in this thesis will be presented in the subsequent background section.

Economic MPC

During the last years, the paradigm to track an a priori given behavior has been diversified, since for many real world applications and problems it might not be clear how to determine what the best operating behavior is. This is often the case for problems whose goal is to optimize the economics of a system. Examples for this are manifold, e.g., maximizing the profit of a chemical plant, minimizing the energy consumption of an air conditioning, maximizing the energy efficiency of renewable power sources, etc. A first step to approach this problem is to use a control structure composed of two layers. While a dynamic real time optimizer (RTO) calculates the optimal steady-states with respect to the economic objective in the top layer (often on a slower time-scale), a stabilizing MPC is employed in the bottom layer to track this reference (see, e.g., Backx et al. (2000); Engell (2007); Jamaludin and Swartz (2015); Kadam and Marquardt (2007); Pang et al. (2015); Würth et al. (2011)). This is advantageous from a computational point of view if the RTO needs to be solved only on a slower time-scale. However, if the economic environmental conditions change frequently, the RTO needs to be solved on a time-scale comparable to the system dynamics. Moreover, nonlinearities can result in a complex optimal operating regime which provides better performance than any operation at an equilibrium.

We have discussed before that an objective is considered within the optimal control problem underlying the MPC and one might wonder why not to choose this objective directly with respect to the desired "economic" goal. This is the idea of *economic model predictive control* (Angeli et al., 2009, 2012; Ellis et al., 2014; Müller and Allgöwer, 2017; Rawlings and Amrit, 2009; Rawlings et al., 2012). It was shown that this approach can provide better performance than the two-layer approach (see, e.g., Angeli et al. (2012); Rawlings and Amrit (2009)).

In fact, since the objective is associated with the economic goal rather than with the stabilization of a set-point, the resulting closed loop can exhibit non-converging behavior, for example periodic or even chaotic behavior (see, e.g., Angeli et al. (2009, 2011); Rawlings et al. (2012)). To this end, it is interesting to investigate whether, and under which conditions, a priori statements about the optimal operating behavior can be made. Of particular interest are optimal operation at steady-state (see, e.g., Angeli et al. (2012); Müller (2014); Müller et al. (2015a)) and optimal periodic operation (see, e.g., Müller and Grüne (2016); Zanon et al. (2016b)). This means that there exists no feasible closed-loop trajectory of the system which results in a better performance than the best steady-state or periodic trajectory, respectively.

Due to the flexibility in the stage cost function and its ability to account for the economic objective directly, economic MPC has been applied to or proposed for a broad range of applications, for example in the process industry (Angeli et al., 2011; Liu et al., 2015; Seban et al., 2016; Tian et al., 2016), for water networks (Grosso et al., 2016; Limón et al., 2014; Revollar et al., 2016; Wang et al., 2016; Zeng and Liu, 2015), for building climate control (Ma and Gupta, 2012; Ma et al., 2012; Mai and Chung, 2015; Patel et al., 2016; Staino et al., 2016), for power grids (Jørgensen et al., 2016; Sokoler et al., 2015; Tedesco et al., 2016; Zachar and Daoutidis, 2016; Zong et al., 2017), for automotive applications (Alrifaee et al., 2015; Halvgaard et al., 2012; Puig et al., 2016; Zhu et al., 2016), for management and planning purposes (Chu et al., 2012; Sokoler et al., 2015; Subramanian et al., 2014), and in the field of power production (Bø and Johansen, 2017; Broomhead et al., 2017; Gros, 2013; Lucia and Engell, 2014; Shafiei et al., 2013; Shaltout et al., 2016).

In the next section, we illustrate why considering disturbances explicitly in the framework of economic MPC can be beneficial from a performance point of view. Moreover, we review some of the results that are available at the interface between robust MPC and economic MPC.

1.2 Motivation for robust economic MPC

In stabilizing MPC, the desired behavior—e.g., controlling a system to a certain state or following a given trajectory—is stated a priori and the stage cost function is designed to achieve this desired behavior. The stage cost function is not the immediate goal of the controller, but a feasible design choice in order to achieve the desired stabilizing behavior. This is in particular contrast to economic MPC, where minimizing the economic objective is the immediate goal of the controller. In a stabilizing setup, the desired behavior of the system is known explicitly beforehand and, thus, stability properties of the closed loop with respect to this behavior shall be guaranteed under disturbances. Computing a controller for all possible disturbances a priori as well as optimizing over general feedback laws is in general computationally intractable (see the discussion in Section 2.2). Therefore, a common approach in robust stabilizing MPC is to only consider an artificial nominal system, which is given by setting all disturbances to zero. In the optimal control problem,

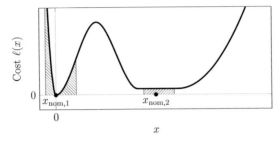

Figure 1.2: Exemplary positive definite stage cost function motivating shift of operational area under disturbances. The two shaded areas represent two sets of states centered at $x_{\mathrm{nom},1}$ and $x_{\mathrm{nom},2}$, respectively, with same cross section highlighting the potential influence of a disturbance on the cost.

the nominal system is considered and replaces the real, disturbance affected system. This problem is then designed in order to satisfy the desired behavior with the nominal system. Obviously, this procedure will in the general case not guarantee that the real system exactly satisfies the desired goal. However, under an appropriate choice of the controller applied to the real system, it can be guaranteed that the real system remains in a set centered at the states of the nominal system. The associated controller should be chosen such that statements on this resulting set can be derived and that it desirably makes the resulting set as small as possible. Considering these aspects, the outlined approach is appropriate when a system shall be stabilized with respect to some desired behavior under disturbance.

As mentioned above, the immediate goal in economic MPC is the optimization of the system's performance given by some arbitrary stage cost function. For this setup, the desired goal need not be representable by stabilizing an a priori determined behavior, and thus, it is not clear if only considering the nominal behavior leads to a good performance of the real system. This could be due to several reasons:

- Disturbances can be beneficial for reaching a "better" state configuration, and thus, might actually help to improve the performance.

- Considering only the nominal system can lead to overlooking states with a poor performance in a close neighborhood of the nominal system. However, the disturbances can cause the real system to drift into these states.

- Applying algorithmic approaches developed for the robust stabilizing case might counteract all disturbances, also those disturbances which are beneficial. Therefore, they can result in a poor performance.

To highlight one of the aspects mentioned before, we have a look at the exemplary stage cost $\ell(x)$ depicted in Figure 1.2. This stage cost function is positive definite with respect to the origin. We assume two systems to be given at the two equilibria $x_{\mathrm{nom},1}$ and $x_{\mathrm{nom},2}$, respectively. In the nominal case, i.e., if no disturbances act on the systems, both systems stay at their respective equilibrium. However, in the disturbed case, the real system states can deviate from their equilibrium and evolve in the sets indicated by the shaded areas. When only considering the nominal system, the optimal behavior for minimizing the stage cost is stabilization of the origin. But, as indicated by the shaded area centered at $x_{\text{nom},1}$, disturbances can lead to a significant increase in the cost in a neighborhood of the origin. Hence, dependent on the disturbance size, the objective, and further knowledge about the closed-loop behavior, it might be beneficial to operate the system in the set centered at $x_{\text{nom},2}$. Even though this results in a worse performance in the undisturbed case, the performance of the disturbed system can be better and can benefit from taking the disturbances into account when setting up the underlying optimization problem.

Thus, simply transferring approaches from robust stabilizing MPC to the economic setup might not always lead to the best result when aiming for the best performance under disturbances. Hence, special approaches have to be developed in order to handle disturbances when an arbitrary performance criterion is to be optimized. This will be the focus of the thesis.

Related work

In the still emerging field of economic MPC, robustness aspects and especially the influence of disturbances on the performance have only been marginally considered. In (Angeli et al., 2016; Ferramosca et al., 2017), economic MPC is investigated under changes in the economic criterion. These setups can, for example, account for time-varying parameters in the stage cost function. Robust economic MPC for tracking a nominally optimal solution, possibly under changing economic criteria, is considered in (Broomhead et al., 2014, 2015; D'Jorge et al., 2016; Huang et al., 2012; Pereira et al., 2016). However, none of these approaches takes the influence of the disturbance on the performance result into consideration. A scenario tree based approach for economic MPC is studied in Lucia et al. (2014). Linear systems and linear objectives are considered in Hovgaard et al. (2011) and a min-max economic MPC (without any guarantees) is investigated in Marquez et al. (2014). The robustness of steady-state optimality under disturbed constraints is considered in Müller and Allgöwer (2012). An approach based on Markovian switching systems is presented in Sopasakis et al. (2017). This approach is limited to a finite number of disturbances but can handle nonlinear systems. In the related field of operations research, considering disturbances in an economic framework is more established aiming for the best result under possible disturbances. To this end, ideas from robust optimization are applied to economic objectives, see, e.g., (Bertsimas and Sim, 2004; Bertsimas et al., 2004, 2011; Beyer and Sendhoff, 2007). However, these approaches focus mainly on stationary solutions. In conclusion, most of the literature in robust economic MPC aims at robust tracking of the optimal behavior of the undisturbed system which, in general, might not be optimal for disturbed systems.

The goal of this thesis is to present different robust economic MPC approaches to incorporate knowledge about the disturbance into the design of the MPC controller in order to achieve a better performance, as well as to deduce statements about performance, optimality, and stability. We present different approaches dependent on the problem structure, the computational complexity, the available information about the disturbance, and how this information is taken into account when designing the MPC controller. Moreover, we provide a thorough theoretical analysis of the presented approaches, investigating and focusing on performance, optimality, as well as on stability.

1.3 Contributions and outline of this thesis

In this section, we present the outline as well as the contributions of this thesis in detail.

Chapter 2 – Background   In this chapter, we provide a basic overview of existing approaches and previous results on stabilizing MPC, robust stabilizing MPC, as well as economic MPC. This chapter provides the background knowledge which the main contributions of the thesis are based on.

Chapter 3 – Min-max robust economic MPC   In this chapter, we present a first approach to handle disturbances in the framework of economic MPC for nonlinear discrete-time systems. The main idea is to adapt approaches from min-max robust stabilizing MPC to the framework of economic MPC and to provide bounds on the asymptotic average performance. We introduce two conceptually different ideas. The first method is adapted from the classical idea of min-max control, that is, one minimizes over the input for the worst-case disturbance sequence. Many of the conceptual findings in min-max robust stabilizing MPC can be retrieved for the presented economic counterpart. In order to reduce the computational complexity of this approach a second idea for min-max robust economic MPC is presented, in which the maximization over disturbance sequences is shifted to a stage-wise maximization over all possible disturbances. However, this approach is more conservative and, in order to reduce conservatism, an adaption for linear systems is presented. For all approaches, constructive methods to find terminal cost functions are presented.

In summary, the main contributions of this chapter are:

- We develop two approaches for handling disturbances in economic MPC within a min-max framework, differing in the way the maximization is considered.

- We provide bounds on the asymptotic average performance of the closed loop when applying the proposed min-max robust economic MPC approaches.

- We illustrate the difference of the two approaches and the validity of their performance bounds by means of numerical examples.

The results in this chapter are based on Bayer et al. (2016b).

Chapter 4 – Tube-based robust economic MPC   In this chapter, we present a second approach to handle disturbances in economic MPC for nonlinear discrete-time systems. The main idea is to consider all possible disturbances specified by an invariant error set by integrating the stage cost function over the respective error set.

We provide a bound on the asymptotic average performance. Based on dissipativity statements adapted to the robust setting, results for optimal operating regimes are obtained for this approach. It turns out that the way of handling information about the disturbance in the underlying optimal control problem plays a significant role for the actual optimal operating regime. Moreover, the derived robust notion of dissipativity turns out to be not only sufficient for optimality, but—under a certain controllability assumption—also a necessary condition. We derive results on asymptotic stability for the considered approach by taking a robust counterpart to strict dissipativity into account.

Additional asymptotic average constraints, that is, constraints that need not be satisfied at each time instance but only on average, are taken into account. Constructive methods are presented, which allow to handle these average constraints in our framework.

To sum up, the main contributions of Chapter 4 are:

- We present a tube-based robust economic MPC approach which considers the disturbances by integration over appropriate invariant error sets and provide a bound on the asymptotic average performance for the proposed approach.

- We investigate optimal operating regimes by means of robust dissipativity results and show sufficiency as well as (under a certain controllability assumption) necessity of this notion for optimal operation at steady-state. By means of a robust notion of strict dissipativity, we prove asymptotic stability of the proposed tube-based robust economic MPC scheme.

- We show how to guarantee satisfaction of asymptotic average constraints in the robust economic MPC framework.

- We illustrate the different findings by numerical examples.

The results in this chapter are based on (Bayer and Allgöwer, 2014; Bayer et al., 2014a,b, 2015b, 2017).

Chapter 5 – Improving robust economic MPC using stochastic information
In this chapter, we consider linear discrete-time systems with additional information on the disturbance, namely stochastic information. Here, the information is given in form of probability density functions. Thus, the expected value of the stage cost function is considered within the optimal control problem. Again, we present bounds on the performance, in this case on the expected value of the asymptotic average performance. Moreover, a constructive approach for finding an appropriate terminal cost function is provided. Optimal operating regimes are investigated by means of stochastic counterparts of the notion of dissipativity. As for the tube-based approach in Chapter 4, sufficiency and (under a certain controllability and convergence assumption) necessity of stochastic dissipativity are proven for the respective notion of optimal operation at steady-state. Additionally, a computationally simpler approach is presented by directly associating the approach from Chapter 4 with distributions over the invariant error sets.

In summary, the main contributions in Chapter 5 are:

- We present two approaches for robust economic MPC which take stochastic information into account.

- We determine bounds on the (expected) asymptotic average performance.

- We prove sufficiency and (under a certain controllability and convergence assumption) necessity of stochastic dissipativity for optimal operation at steady-state in the respective stochastic notion.

The results in this chapter are based on (Bayer et al., 2015a, 2016a, 2017).

Chapter 6 – Comparison of the approaches   In this chapter, we compare the different approaches as well as the different notions of optimal operation at steady-state which have been developed in the previous chapters. In addition to the different optimal operating regimes in the previous chapters, another notion of optimality is presented depending only on the dynamics, the stage cost, and the constraints. This notion is inspired by the nominal setup, and it is in particular independent of the MPC algorithm considered. Again, sufficiency and (under a certain controllability assumption) necessity of the respective dissipativity notion for steady-state optimality are proven. To sum up, the different notions of optimality as well as the different robust economic MPC approaches are contrasted with each other in numerical examples.

The main contributions of this chapter are:

- We present a new notion of robust optimal operation at steady-state which only depends on the dynamics, the stage cost, and the constraints. We prove sufficiency and (under a certain controllability assumption) necessity of the respective dissipativity notion for steady-state optimality.

- We investigate and discuss the different notions of optimal steady-state operation and contrast them with each other by means of a numerical example.

- We present a numerical example with a detailed discussion for all presented robust economic MPC approaches.

The results in this chapter are based on Bayer et al. (2017), the numerical example was previously considered in (Bayer et al., 2014b, 2015a, 2016a,b).

Chapter 7 – Conclusions   In this chapter, we summarize the contributions of this thesis and discuss some open future research topics.

Appendix A provides the technical proofs for the results in this thesis.

All numerical simulations in this thesis were implemented in MATLAB R2015b using the MPT 2.6.2 (Kvasnica et al., 2004). The optimizations were executed with the interior-point algorithm of `fmincon` (MATLAB, 2015).

Chapter 2

Background

After having motivated the idea of robust economic MPC in the previous chapter, in this chapter we want to focus on and recapitulate the most important results in the fields of stabilizing—and especially robust stabilizing—MPC, as well as of economic MPC.

2.1 Stabilizing MPC

In this section, we consider nominal MPC schemes, that is, we assume the model employed within the prediction to be exact. Thus, no plant/model-mismatch or external disturbances are assumed. This leads to the consideration of nonlinear discrete-time, time-invariant systems of the form

$$x(t+1) = f\big(x(t), u(t)\big), \qquad x(0) = x_0, \tag{2.1}$$

where $f : \mathbb{R}^n \times \mathbb{R}^m \to \mathbb{R}^n$, $x(t) \in \mathbb{X} \subseteq \mathbb{R}^n$ is the system state and $u(t) \in \mathbb{U} \subseteq \mathbb{R}^m$ is the input to the system at time $t \in \mathbb{I}_{\geq 0}$. The initial state is denoted by $x_0 \in \mathbb{X}$. As constraints, we have given pointwise-in-time constraints of the form

$$\big(x(t), u(t)\big) \in \mathbb{Z} \subseteq \mathbb{X} \times \mathbb{U}, \tag{2.2}$$

for all $t \in \mathbb{I}_{\geq 0}$, such that mixed state and input constraints can be considered. The following standard assumptions are imposed on the dynamics and the constraints:

Assumption 2.1. *The function f is continuous.*

Assumption 2.2. *The set $\mathbb{Z} \subseteq \mathbb{X} \times \mathbb{U}$ is compact.[1]*

The main idea in stabilizing MPC is to stabilize a given system at a setpoint, a set, or a trajectory while the constraints on the state and/or the input must be satisfied at all time. We use the standard definition of stability according to (Jiang and Wang, 2002; Rawlings and Mayne, 2009):

Definition 2.1 (Stability (constrained)). *Suppose $\mathbb{X} \subseteq \mathbb{R}^n$ is positively invariant[2] for the system*

$$x(t+1) = g\big(x(t)\big), \qquad x(0) = x_0, \tag{2.3}$$

---

[1]For the results in (robust) stabilizing MPC, it would be sufficient to assume that $\mathbb{U}$ is compact and $\mathbb{Z}$ is closed. However, compactness of $\mathbb{Z}$ required for the results on (robust) economic MPC and, thus, assumed throughout the thesis.

[2]A set $\mathcal{A}$ is *positively invariant* for system (2.3) if for each $x \in \mathcal{A}$, it holds that $g(x) \in \mathcal{A}$.

where $g : \mathbb{R}^n \to \mathbb{R}^n$. *A closed and positively invariant set* $\mathcal{A} \subseteq \mathbb{X}$ *is stable for* (2.3) *if, for every* $\varepsilon > 0$, *there exists a* $\delta > 0$ *such that* $|x_0|_{\mathcal{A}} \leq \delta$ *and* $x_0 \in \mathbb{X}$ *imply* $|x(t)|_{\mathcal{A}} \leq \varepsilon$ *for all* $t \in \mathbb{I}_{\geq 0}$. *It is* asymptotically stable *with region of attraction* $\mathbb{X}$ *if it is stable and* $\lim_{t\to\infty} |x(t)|_{\mathcal{A}} = 0$ *for all* $x_0 \in \mathbb{X}$.[3]

In case g is continuous and $\mathcal{A}$ is a compact set, it was shown in Jiang and Wang (2002) that asymptotic stability (Definition 2.1) is equivalent to the existence of a $\mathcal{KL}$ function β such that for all $x_0 \in \mathbb{X}$ it holds that $|x(t)|_{\mathcal{A}} \leq \beta(|x_0|_{\mathcal{A}}, t)$ for all $t \in \mathbb{I}_{\geq 0}$.

For the remainder of this section, we restrict ourselves to the stabilization of a given setpoint only, that is, $\mathcal{A} = \{x_s\}$. Given the dynamics and the constraints, the objective is stated as follows: We want to stabilize system (2.1) at a setpoint x_s, for which we assume that there exists an input u_s such that (x_s, u_s) is a feasible equilibrium point, that is, $x_s = f(x_s, u_s)$ and $(x_s, u_s) \in \mathbb{Z}$. Moreover, the pointwise-in-time constraints (2.2) must be satisfied at each time $t \in \mathbb{I}_{\geq 0}$.

In the considered nominal case, the receding horizon control law is determined by solving the following optimization problem at each time $t \in \mathbb{I}_{\geq 0}$ and for a given $x(t)$:

Problem 2.2.

$$\underset{\boldsymbol{u}(t)}{\text{minimize }} J_N\Big(x(t), \boldsymbol{u}(t)\Big)$$

subject to

$$x(k+1|t) = f\Big(x(k|t), u(k|t)\Big), \qquad \forall k \in \mathbb{I}_{[0,N-1]}, \qquad (2.4\text{a})$$

$$x(0|t) = x(t), \qquad (2.4\text{b})$$

$$\Big(x(k|t), u(k|t)\Big) \in \mathbb{Z}, \qquad \forall k \in \mathbb{I}_{[0,N-1]}, \qquad (2.4\text{c})$$

$$x(N|t) \in \mathbb{X}_{\mathrm{f}}, \qquad (2.4\text{d})$$

where

$$J_N\Big(x(t), \boldsymbol{u}(t)\Big) = \sum_{k=0}^{N-1} \ell\Big(x(k|t), u(k|t)\Big) + V_{\mathrm{f}}\Big(x(N|t)\Big). \qquad (2.5)$$

In the stated problem, $\boldsymbol{u}(t) = \{u(0|t), \ldots, u(N-1|t)\}$ denotes the predicted input sequence and, by $\boldsymbol{x}(t) = \{x(0|t), \ldots, x(N|t)\}$, the associated predicted state sequence is represented, where $N \in \mathbb{I}_{>0}$ is the prediction horizon. The notation $u(k|t)$ expresses a k-step ahead prediction, predicted at time t. For the terminal region it holds $\mathbb{X}_{\mathrm{f}} \subseteq \mathbb{X}$, with $\mathbb{X}_{\mathrm{f}}$ being a closed set. The stage cost and the terminal cost are given by $\ell : \mathbb{X} \times \mathbb{U} \to \mathbb{R}$ and $V_{\mathrm{f}} : \mathbb{X}_{\mathrm{f}} \to \mathbb{R}$, respectively, for which the following assumptions are imposed:

Assumption 2.3. *The stage cost ℓ is continuous.*

Assumption 2.4. *The terminal cost V_{f} is continuous.*

A solution to Problem 2.2 exists (see, e.g., Rawlings and Mayne (2009, Proposition 2.4)) and we denote the minimizing input sequence[4] by $\boldsymbol{u}^*(t) = \{u^*(0|t), \ldots, u^*(N-1|t)\}$, and the associated state sequence by $\boldsymbol{x}^*(t) = \{x^*(0|t), \ldots, x^*(N|t)\}$. The value function is $V_N(x(t)) = J_N(x(t), \boldsymbol{u}^*(t))$. We define the controller by the following algorithm:

---

[3] A set $\mathcal{A}$ is *globally asymptotically stable*, if it is asymptotically stable with region of attraction $\mathbb{X} = \mathbb{R}^n$.
[4] For ease of presentation, we assume $\boldsymbol{u}^*(t)$ to be unique. If this is not the case, one minimizer can be assigned. This consideration holds for all optimal control problems investigated in the remainder of the thesis.

Algorithm 2.3 (Stabilizing model predictive control). *At each time instant* $t \in \mathbb{I}_{\geq 0}$, *measure the state* $x(t)$ *and solve Problem 2.2. Apply the control input* $u(t) := u^*(0|t)$ *to system* (2.1).

Under certain assumptions on the stage cost ℓ, the terminal cost V_f, and the terminal region $\mathbb{X}_f$, one can show that the setpoint x_s is asymptotically stable for the closed-loop system

$$x(t+1) = f\big(x(t), u^*(0|t)\big). \tag{2.6}$$

The following assumptions offer one way to do so:

Assumption 2.5. *The stage cost function satisfies* $\ell(x_s, u_s) = 0$ *and there exists a class* $\mathcal{K}_\infty$ *function* α_1 *such that* $\ell(x, u) \geq \alpha_1(|x - x_s|)$ *for all* $(x, u) \in \mathbb{Z}$. *The terminal cost function satisfies* $V_f(x_s) = 0$ *and* $V_f(x) \geq 0$ *for all* $x \in \mathbb{X}_f$.

Assumption 2.6. *The terminal region* $\mathbb{X}_f \subseteq \mathbb{X}$ *is closed and contains* x_s *in its interior. There exists a local auxiliary controller* $\kappa_f : \mathbb{X}_f \to \mathbb{U}$ *such that for all* $x \in \mathbb{X}_f$ *it holds that*

(i) $\big(x, \kappa_f(x)\big) \in \mathbb{Z}$,

(ii) $f\big(x, \kappa_f(x)\big) \in \mathbb{X}_f$, *and*

(iii) $V_f\big(f(x, \kappa_f(x))\big) - V_f(x) \leq -\ell\big(x, \kappa_f(x)\big) + \ell(x_s, u_s)$.

Condition (i) of Assumption 2.6 guarantees feasibility of the terminal auxiliary controller κ_f on the whole terminal region $\mathbb{X}_f$, while (ii) guarantees positive invariance of the terminal region $\mathbb{X}_f$ under κ_f. Together with Assumption 2.5, Assumption 2.6 (iii) states that the terminal cost V_f can be employed as a control Lyapunov function on the terminal region. The set of all states x for which there is a solution for Problem 2.2 with $x(t) = x$ is denoted by $\mathbb{X}_N := \{x \in \mathbb{R}^n : \exists \boldsymbol{u} = \{u(0), \ldots, u(N-1)\} \text{ s.t. } x(0) = x, x(k+1) = f(x(k), u(k)), (x(k), u(k)) \in \mathbb{Z}, \forall k \in \mathbb{I}_{[0, N-1]}, \text{ and } x(N) \in \mathbb{X}_f\}$.

Theorem 2.4. *Let* $x_0 \in \mathbb{X}_N$. *Suppose that Assumptions 2.1–2.6 are satisfied. Then, the closed-loop system* (2.6) *resulting from applying Algorithm 2.3 satisfies the pointwise-in-time constraints* (2.2) *for all* $t \in \mathbb{I}_{\geq 0}$ *and Problem 2.2 is feasible for all* $t \in \mathbb{I}_{\geq 0}$. *Furthermore,* x_s *is an asymptotically stable equilibrium with region of attraction* $\mathbb{X}_N$.

This theorem and its proof can, for example, be found in (Grüne and Pannek, 2011; Mayne et al., 2000; Rawlings and Mayne, 2009). A continuous-time version of this theorem also exists, see, e.g., (Chen and Allgöwer, 1998; Findeisen et al., 2003; Fontes, 2001). Constructive approaches to find an appropriate terminal auxiliary controller, terminal region, and terminal cost can be found in Rawlings and Mayne (2009) for the discrete-time setup, and in Chen and Allgöwer (1998) in case of continuous time. Other approaches to provide asymptotic stability, which are independent of the terminal cost function and/or the terminal region, are presented, for example, in (Grüne, 2013; Grüne and Stieler, 2014; Reble and Allgöwer, 2012; Reble et al., 2012). These approaches are based on a certain controllability assumption but are beyond the scope of this thesis.

2.2 Robust stabilizing MPC

In robust model predictive control, we are concerned with systems that are uncertain, i.e., their nominal system description used in the prediction does not represent the exact behavior. There are different sources for these mismatches. In robust MPC, one discriminates between uncertainties and disturbances; the first is due to unknown parameters or unmodeled dynamics, the second is caused by external influences on the system. When computing controls based on inaccurate models, the statements provided for stabilizing MPC in Section 2.1 need not be maintained.

In this section, we focus on systems with disturbances. Therefore, we consider system dynamics of the form

$$x(t+1) = f\Big(x(t), u(t), w(t)\Big), \qquad x(0) = x_0, \tag{2.7}$$

where $f : \mathbb{R}^n \times \mathbb{R}^m \times \mathbb{R}^q \to \mathbb{R}^n$. Again, $x(t) \in \mathbb{X} \subseteq \mathbb{R}^n$ is the system state and $u(t) \in \mathbb{U} \subseteq \mathbb{R}^m$ the input to the system. The disturbance $w(t) \in \mathbb{W} \subset \mathbb{R}^q$ represents an uncontrollable external input to the system at time $t \in \mathbb{I}_{\geq 0}$. Although disturbances act on the system, the considered robust MPC schemes are required to satisfy the pointwise-in-time constraint (2.2) for all times $t \in \mathbb{I}_{\geq 0}$.

Due to the unknown disturbance, system (2.7) cannot be stabilized at a single state. In order to be able to provide feasibility and stability results, the disturbance is usually assumed to be bounded. To be more precise, the following assumption is imposed:

Assumption 2.7. *For each $t \in \mathbb{I}_{\geq 0}$, the disturbance satisfies $w(t) \in \mathbb{W} \subset \mathbb{R}^q$, where $\mathbb{W}$ is a compact and convex set containing the origin in its interior.*

In order to be able to derive stability results for the disturbed case, we introduce positive invariance and control invariance under disturbances (Blanchini, 1999; Kerrigan, 2000; Rawlings and Mayne, 2009).

Definition 2.5 (Robust positive invariance (RPI))**.** *Suppose that Assumption 2.7 is satisfied. A set $\mathcal{A} \subseteq \mathbb{R}^n$ is robust positively invariant (RPI) for a system*

$$x(t+1) = g\Big(x(t), w(t)\Big), \qquad x(0) = x_0, \tag{2.8}$$

where $g : \mathbb{R}^n \times \mathbb{R}^q \to \mathbb{R}^n$, if, for all $x \in \mathcal{A}$ and for all $w \in \mathbb{W}$, it holds $g(x, w) \in \mathcal{A}$.

Definition 2.6 (Robust control invariance (RCI))**.** *Suppose that Assumption 2.7 is satisfied. A set $\mathcal{A} \subseteq \mathbb{R}^n$ is robust control invariant (RCI) for system (2.7), if, for all $x \in \mathcal{A}$, there exists a $u \in \mathbb{U}$ with $(x, u) \in \mathbb{Z}$, such that for all $w \in \mathbb{W}$, it holds $f(x, u, w) \in \mathcal{A}$.*

Assumption 2.8. *For system (2.7) there exists a compact RCI set with non-empty interior.*

Remark 2.7. *Compactness does not pose a major restriction on the RCI set since many constructive approaches provide compact sets in any case.*

We denote the solution of (2.8) at time t given the initial state x_0 for some admissible disturbance sequence $\boldsymbol{w} := \{w(0), w(1), \dots\}$ with $\phi(t) := \phi(t; x_0, \boldsymbol{w})$ and the set of all solutions of (2.8) starting at x_0 with $S(x_0)$. With this, we can state stability under disturbances (Jiang and Wang, 2002; Rawlings and Mayne, 2009):

Definition 2.8 (Stability under disturbances). *Suppose that Assumption 2.7 is satisfied and that $\mathbb{X} \subseteq \mathbb{R}^n$ is robust positively invariant for system (2.8). A closed and robust positively invariant set $\mathcal{A} \subseteq \mathbb{X}$ is stable in $\mathbb{X}$ for (2.8) if, for every $\varepsilon > 0$, there exists a $\delta > 0$ such that $|x_0|_{\mathcal{A}} \leq \delta$ and $x_0 \in \mathbb{X}$ imply $|\phi(t)|_{\mathcal{A}} \leq \varepsilon$ for each solution $\phi(\cdot) \in S(x_0)$ and for all $t \in \mathbb{I}_{\geq 0}$. It is asymptotically stable with region of attraction $\mathbb{X}$ if it is stable and each solution $\phi(\cdot) \in S(x_0)$ satisfies $\lim_{t \to \infty} |\phi(t)|_{\mathcal{A}} = 0$ for all $x_0 \in \mathbb{X}$.*

As in the nominal case, an alternative but equivalent definition of asymptotic stability of $\mathcal{A}$ can be provided in case $\mathcal{A}$ is compact and g is continuous, namely existence of a $\mathcal{KL}$ function β such that for each $x_0 \in \mathbb{X}$ each solution $\phi(\cdot) \in S(x_0)$ satisfies $|\phi(t)|_{\mathcal{A}} \leq \beta(|x_0|_{\mathcal{A}}, t)$ for all $t \in \mathbb{I}_{\geq 0}$.

Another conceptual idea in robust stabilizing MPC is *input-to-state stability* (ISS) (Jiang and Wang, 2001; Sontag, 1989; Sontag and Wang, 1995) where one seeks a uniform gain on the disturbance sequence $\boldsymbol{w}$ in order to provide a bound on the state sequence. The ℓ_∞-norm of the sequence $\boldsymbol{w}$ is denoted by $\|\boldsymbol{w}\| := sup_{t \geq 0} |w(t)|$. In order to account for the constraints usually considered in MPC, we use a regional version of ISS (Magni et al., 2006; Raimondo et al., 2009):

Definition 2.9 (Regional input-to-state stability (ISS)). *Suppose that Assumption 2.7 is satisfied and $\mathbb{X} \subseteq \mathbb{R}^n$ is robust positively invariant for (2.8). System (2.8) is input-to-state stable (ISS) in $\mathbb{X}$ if there exist a $\mathcal{KL}$ function β and a $\mathcal{K}$ function σ such that, for each $x_0 \in \mathbb{X}$ and each admissible disturbance sequence $\boldsymbol{w}$, it holds that*

$$\left| \phi(t; x_0, \boldsymbol{w}_t) \right| \leq \beta\left(|x_0|, t \right) + \sigma\left(\|\boldsymbol{w}_t\| \right) \tag{2.9}$$

for all $t \in \mathbb{I}_{\geq 0}$, where $\phi(t; x_0, \boldsymbol{w}_t)$ is the solution of (2.8) at time t given the initial state x_0 and the disturbance sequence $\boldsymbol{w}_t := \{w(0), \ldots, w(t-1)\}$.

Input-to-state stability is closely related to stability (Definition 2.1) since the definition implies asymptotic stability if the disturbance sequence is identically zero.

Even though nominal MPC algorithms can provide a certain robustness against disturbances (Allan et al., 2016; Grimm et al., 2007; Pannocchia et al., 2011a,b; Yu et al., 2011, 2014), there is no guarantee that robustness is always provided by MPC (Grimm et al., 2004; Teel, 2004). Thus, it is reasonable to structurally incorporate the disturbances—or the knowledge which is provided—within the design of the model predictive controller; on the one hand to satisfy the constraints, on the other hand to provide robust stability results.

When disturbances are present, there is a significant difference between open-loop control and feedback control. In the first case one only optimizes for input values in the finite horizon optimal control problem; in the second case, one optimizes for general feedback policies (see, e.g., Rawlings and Mayne (2009)). Within the available robust stabilizing MPC approaches there is a significant difference in determining the control. However, all of them have in common that if Assumption 2.7 is satisfied only some set can be robustly stabilized or some ISS statement can be derived. This will be further investigated subsequently for *min-max robust stabilizing MPC* and *tube-based robust stabilizing MPC*.

2.2.1 Min-max robust stabilizing MPC

In min-max robust stabilizing MPC, the disturbances are explicitly taken into account within the optimization problem in order to consider their influence. The optimization problem is adapted by first maximizing over all possible disturbances and then second minimizing over the feedback control. From an intuitive point of view, the optimization problem takes the worst case into account which could be caused with the disturbance within the prediction and optimizes the feedback control accordingly.

In the most general case of min-max robust stabilizing MPC, one can optimize over general feedback control laws of the form $\boldsymbol{\pi}_t = \{\pi_{0|t}(\cdot), \ldots, \pi_{N-1|t}(\cdot)\}$, where $\pi_{k|t} : \mathbb{X} \to \mathbb{U}$, $k \in \mathbb{I}_{[0,N-1]}$, and $u(t) := \pi_{0|t}(x(t))$ is the control action applied to the system at time $t \in \mathbb{I}_{\geq 0}$ (Kerrigan and Maciejowski, 2004). Given a state $x(t)$, a sequence of state feedback control laws $\boldsymbol{\pi}_t$, and a disturbance sequence $\boldsymbol{w}_N = \{w(0), \ldots, w(N-1)\} \in \mathbb{W}^N$, we denote the solution to (2.7), i.e., $z(k+1) = f(z(k), \pi_{k|t}(z(k)), w(k))$, $z(0) = x(t)$, $k \in \mathbb{I}_{[0,N-1]}$, by $\phi(k; x(t), \boldsymbol{\pi}_t, \boldsymbol{w}_N)$, $k \in \mathbb{I}_{[0,N]}$.

Even though this idea is in most cases computationally intractable since the optimization problem results in an infinite dimensional optimization problem, it has some appealing properties. We present this most general version as a starting point for the subsequent analysis and gradually introduce further restrictions. The min-max optimization problem to be solved in order to determine the receding horizon control law at each time $t \in \mathbb{I}_{\geq 0}$ for a given $x(t)$ is as follows:

Problem 2.10.

$$\underset{\boldsymbol{\pi}_t}{\text{minimize}} \ \underset{\boldsymbol{w}(t)}{\text{maximize}} \ J_N\big(x(t), \boldsymbol{\pi}_t, \boldsymbol{w}(t)\big)$$

subject to

$$\Big(\phi(k; x(t), \boldsymbol{\pi}_t, \boldsymbol{w}_N), \pi_{k|t}\big(\phi(k; x(t), \boldsymbol{\pi}_t, \boldsymbol{w}_N)\big)\Big) \in \mathbb{Z}, \quad \forall k \in \mathbb{I}_{[0,N-1]}, \forall \boldsymbol{w}_N \in \mathbb{W}^N, \quad (2.10a)$$

$$\phi(N; x(t), \boldsymbol{\pi}_t, \boldsymbol{w}_N) \in \mathbb{X}_\mathrm{f}, \qquad\qquad \forall \boldsymbol{w}_N \in \mathbb{W}^N, \quad (2.10b)$$

$$w(k|t) \in \mathbb{W}, \quad \forall k \in \mathbb{I}_{[0,N-1]}, \qquad\qquad (2.10c)$$

where

$$J_N\big(x(t), \boldsymbol{\pi}_t, \boldsymbol{w}(t)\big) = \sum_{k=0}^{N-1} \ell\big(x(k|t), u(k|t)\big) + V_\mathrm{f}\big(x(N|t)\big), \qquad (2.11)$$

with the definitions $\boldsymbol{w}(t) = \{w(0|t), \ldots, w(N-1|t)\}$, $x(k|t) := \phi(k; x(t), \boldsymbol{\pi}_t, \boldsymbol{w}(t))$, $k \in \mathbb{I}_{[0,N]}$, *and* $u(k|t) := \pi_{k|t}\big(\phi(k; x(t), \boldsymbol{\pi}_t, \boldsymbol{w}(t))\big)$, $k \in \mathbb{I}_{[0,N-1]}$.

We note that we do not further specify the terminal region $\mathbb{X}_\mathrm{f} \subseteq \mathbb{X}$ and the terminal cost $V_\mathrm{f} : \mathbb{R}^n \to \mathbb{R}$. Depending on the stability result to be derived, different assumptions can be posed on these ingredients of the optimization problem.

Remark 2.11. *We note that the constraints* (2.10a) *and* (2.10b) *have to hold for* all *possible disturbance sequences* $\boldsymbol{w}_N \in \mathbb{W}^N$ *in order to be able to guarantee recursive feasibility, whereas the objective* (2.11) *is given with respect to the worst-case disturbance* $\boldsymbol{w}(t)$ *only, which is an optimization variable.*

In the min-max setup, the controller is defined by the following algorithm:

Algorithm 2.12 (Stabilizing min-max robust model predictive control). *At each time instant* $t \in \mathbb{I}_{t \geq 0}$, *measure the state* $x(t)$ *and solve Problem 2.10. Apply the control input* $u(t) := \pi_{0|t}^*(x(t))$ *to system* (2.7).

As mentioned above, this general optimization problem is computationally intractable. Thus, different concepts have been presented to overcome this deficiency. A common approach is to parametrize the feedback control law in order to derive an optimization problem with finitely many optimization variables. These include affine parametrizations (see, e.g., Goulart et al. (2006); Löfberg (2003)) or arbitrary input parametrizations (see, e.g., Limón et al. (2009)). Other ideas exploit structural characteristics of the optimal input for special problem setups (see, e.g., Bemporad et al. (2003); Ramirez and Camacho (2003)), employ an $\mathcal{H}_\infty$-type controller design method (Raimondo et al., 2009), or approximate the worst case appropriately in order to reformulate the problem as a minimization only (Raimondo et al., 2007). For linear systems with polytopic disturbances and convex stage cost function, one can utilize a combinatorial approach considering the combination of all disturbance vertices along the prediction horizon (see, e.g., Kerrigan and Maciejowski (2004); Scokaert and Mayne (1998)). Instead of solving Problem 2.10 directly, one can also reformulate the optimization problem by means of dynamic programming and exploit the structure of the resultant problem (Bemporad et al., 2003; Lazar et al., 2008; Lee and Yu, 1997; Limón et al., 2006; Mayne, 2001; Rawlings and Mayne, 2009).

Considering stability in min-max robust stabilizing MPC for the closed-loop system resulting from applying Algorithm 2.12, one cannot guarantee that a steady-state is asymptotically stable as unknown disturbances affect the system at each time instance. Within the literature, the two notions presented before—asymptotic stability of an invariant set (see, e.g., Kerrigan and Maciejowski (2004); Rawlings and Mayne (2009)) and input-to-state stability (see, e.g., Lazar et al. (2008); Limón et al. (2009); Magni et al. (2006); Raimondo et al. (2009))—are investigated in min-max robust stabilizing MPC.[5]

2.2.2 Tube-based robust stabilizing MPC

While feedback control is superior to open-loop control whenever disturbances act on the system, the computational complexity to optimize for arbitrary control laws is impractical. However, for the undisturbed system, feedback control and open-loop control are equivalent (Rawlings and Mayne, 2009). Hence, a common idea in robust (stabilizing) MPC is to parametrize the feedback by calculating an open-loop control for the artificial undisturbed system and account for the disturbances acting on the real system by designing an additional static feedback policy.

The nominal system is given by

$$z(t+1) = f\big(z(t), v(t), 0\big), \qquad z(0) = z_0, \tag{2.12}$$

where $f : \mathbb{R}^n \times \mathbb{R}^m \times \mathbb{R}^q \to \mathbb{R}^n$ is the same system dynamics as in (2.7). The nominal state and input at time $t \in \mathbb{I}_{\geq 0}$ are denoted by $z(t) \in \mathbb{X}$ and $v(t) \in \mathbb{R}^m$, respectively; the initial state is given by $z_0 \in \mathbb{X}$.

In order to relate the real with the artificial nominal system, one can consider the error

$$e(t) := x(t) - z(t). \tag{2.13}$$

---

[5]In the literature, min-max MPC schemes can as well be used in an $\mathcal{H}_\infty$ framework by using a disturbance-dependent stage cost function $\ell(x, u, w)$, e.g., $\ell(x, u) - \ell_w(w)$ (cf., Chen et al. (1997); Magni and Scattolini (2007); Rawlings and Mayne (2009)). This can be particularly useful when the disturbances are state dependent, for example, when they represent uncertainties. However, we will not consider this approach in the subsequent sections.

Parameterizing the real input by the general static continuous feedback

$$u(t) = \varphi\big(v(t), x(t), z(t)\big), \tag{2.14}$$

where $\varphi : \mathbb{R}^m \times \mathbb{R}^n \times \mathbb{R}^n \to \mathbb{R}^m$, the error dynamics are

$$e(t+1) = f\big(x(t), \varphi(v(t), x(t), z(t)), w(t)\big) - f\big(z(t), v(t), 0\big), \qquad e(0) = x_0 - z_0. \tag{2.15}$$

Since the open-loop control is determined for the nominal system only, one has to consider the error in order to account for the neglected parts of the real system. By taking the errors appropriately into account within the problem setup, one aims to guarantee recursive feasibility as well as (robust) stabilization of a set even though only taking the nominal system into account.

One of the most commonly applied ways to consider the influence of the disturbance is by using RCI sets. Therefore, we introduce a new definition of RCI sets for the error dynamics (2.15):

Definition 2.13 (Robust control invariant error set). *Suppose that Assumption 2.7 is satisfied. A set $\Omega \subseteq \mathbb{R}^n$ is robust control invariant (RCI) for the error dynamics (2.15) if there exists a feedback control law $u(t) = \varphi(v(t), x(t), z(t))$ such that for all $x(t), z(t) \in \mathbb{R}^n$ and $v(t) \in \mathbb{R}^m$ with $e(t) = x(t) - z(t) \in \Omega$ and $\big(x(t), \varphi(v(t), x(t), z(t))\big) \in \mathbb{Z}$, and for all $w(t) \in \mathbb{W}$, it holds $e(t+1) \in \Omega$.*

We note that this new definition is more conservative than Definition 2.6 because of the special structure of the feedback such that invariance is guaranteed *for all feasible $v(t)$*.

Remark 2.14. *Definition 2.13 is a slight adaptation of the classical definition of robust control invariant sets, see, e.g., Definition 2.6 and (Blanchini, 1999; Kerrigan, 2000). Those sets are defined for systems of the form $e(t+1) = h(e(t), u(t), w(t))$, where $h : \mathbb{R}^n \times \mathbb{R}^m \times \mathbb{R}^q \to \mathbb{R}^n$, and an accompanying feedback of the form $u = k(e)$ such that the set Ω is positively invariant for the corresponding closed-loop system. For general nonlinear systems, the given error system (2.15) cannot necessarily be transferred into this form. Nevertheless, as the conceptual idea is the same, we also refer to the set provided by Definition 2.13 as an RCI set.*

The general idea of tube-based robust MPC underlies Definition 2.13: Since it is impossible to control all possible disturbance sequences, the feedback law is chosen such that *any* real system state stays in a neighborhood of the artificial nominal system. These neighborhoods—which are represented by the RCI error sets—are referred to as tubes and they are centered at the states of the nominal system. Furthermore, the tubes and, hence, all possible real states can be controlled in a bundle by means of controlling the nominal system.

We want to be able to guarantee constraint satisfaction despite the disturbances. Thus, we make use of the tubes just defined and tighten the constraint set $\mathbb{Z}$ accordingly. The tightened set denotes all possible nominal state and input combinations guaranteeing satisfaction of the original constraints and robust invariance of the error in Ω at the same time. This leads to

$$\overline{\mathbb{Z}} := \Big\{ (z, v) \in \mathbb{R}^n \times \mathbb{R}^m : \big(x, \varphi(v, x, z)\big) \in \mathbb{Z} \text{ for all } x \in \{z\} \oplus \Omega \Big\}. \tag{2.16}$$

We denote by $\overline{\mathbb{X}}$ the projection of $\overline{\mathbb{Z}}$ onto $\mathbb{X}$, i.e., $\overline{\mathbb{X}} := \{ z \in \mathbb{X} : \exists v \in \mathbb{R}^m \text{ s.t. } (z, v) \in \overline{\mathbb{Z}} \}$. Moreover, we assume that the disturbances are sufficiently small with respect to the constraints:

Assumption 2.9. *The set $\overline{Z}$ is non-empty and compact.*

With this, the nominal optimization problem to be solved in order to determine the receding horizon control law at each time $t \in \mathbb{I}_{\geq 0}$ for a given $x(t)$ is as follows:

Problem 2.15.

$$\underset{z(0|t),\,\boldsymbol{v}(t)}{\text{minimize}} \ J_N\Big(z(0|t), \boldsymbol{v}(t)\Big)$$

subject to

$$z(k+1|t) = f\Big(z(k|t), v(k|t), 0\Big), \qquad \forall k \in \mathbb{I}_{[0,N-1]}, \tag{2.17a}$$

$$x(t) \in \{z(0|t)\} \oplus \Omega, \tag{2.17b}$$

$$\Big(z(k|t), v(k|t)\Big) \in \overline{Z}, \qquad \forall k \in \mathbb{I}_{[0,N-1]}, \tag{2.17c}$$

$$z(N|t) \in \overline{\mathbb{X}}_{\mathrm{f}}, \tag{2.17d}$$

where

$$J_N\Big(z(0|t), \boldsymbol{v}(t)\Big) = \sum_{k=0}^{N-1} \ell\Big(z(k|t), v(k|t)\Big) + V_{\mathrm{f}}\Big(z(N|t)\Big). \tag{2.18}$$

Similar as in Section 2.1, we denote the minimizing predicted nominal input sequence by $\boldsymbol{v}^*(t) = \{v^*(0|t), \ldots, v^*(N-1|t)\}$, the associated nominal state sequence by $\boldsymbol{z}^*(t) = \{z^*(0|t), \ldots, z^*(N|t)\}$. Again, the terminal region $\overline{\mathbb{X}}_{\mathrm{f}} \subseteq \overline{\mathbb{X}}$ is a closed set and Assumption 2.4 is imposed on the terminal cost $V_{\mathrm{f}} : \overline{\mathbb{X}}_{\mathrm{f}} \to \mathbb{R}$, i.e., it is continuous on $\overline{\mathbb{X}}_{\mathrm{f}}$. The tube-based robust stabilizing controller is defined by the following algorithm:

Algorithm 2.16 (Stabilizing tube-based robust model predictive control). *At each time instant $t \in \mathbb{I}_{\geq 0}$, measure the state $x(t)$ and solve Problem 2.15. Apply the control input $u(t) := \varphi(v^*(0|t), x(t), z^*(0|t))$ to system (2.7).*

Applying Algorithm 2.16 results in the closed-loop system

$$x(t+1) = f\Big(x(t), \varphi(v^*(0|t), x(t), z^*(0|t)), w(t)\Big). \tag{2.19}$$

By $\overline{\mathbb{X}}_N$, we denote the set of all feasible nominal states, i.e., $\overline{\mathbb{X}}_N := \{z \in \overline{\mathbb{X}} : \exists \boldsymbol{v} = \{v(0), \ldots, v(N-1)\} \text{ s.t. } z(0) = z, z(k+1) = f(z(k), v(k), 0), (z(k), v(k)) \in \overline{Z}, \forall k \in \mathbb{I}_{[0,N-1]}, \text{ and } z(N) \in \overline{\mathbb{X}}_{\mathrm{f}}\}$.

Theorem 2.17. *Let $x \in \overline{\mathbb{X}}_N \oplus \Omega$. Suppose that Assumptions 2.1, 2.3–2.5, 2.7, and 2.9 are satisfied and that Assumption 2.6 is satisfied with respect to the nominal system (2.12) and the tightened constrained (2.16). The closed-loop system (2.19) resulting from applying Algorithm 2.16 satisfies the pointwise-in-time constraints (2.2) for all $t \in \mathbb{I}_{\geq 0}$ and Problem 2.15 is feasible for all $t \in \mathbb{I}_{\geq 0}$. Furthermore, the closed-loop system (2.19) converges to $\{x_s\} \oplus \Omega$, that is, $\lim_{t\to\infty} |x(t)|_{\{x_s\}\oplus\Omega} = 0$, with region of attraction $\overline{\mathbb{X}}_N \oplus \Omega$.*

The proof for this result can be found in, e.g., Bayer et al. (2013) as well as (for linear system dynamics) in (Mayne et al., 2005; Rawlings and Mayne, 2009). We note that in Rawlings and Mayne (2009), convergence of the closed-loop system (2.19) is replaced by asymptotic stability, which can be shown for the *composite* system (2.7) and (2.12). This does *not* necessarily induce asymptotic stability of the closed-loop system (2.19) as one can only derive asymptotic convergence but one cannot guarantee stability.[6]

In contrast to Problem 2.2, the nominal initial state is an additional optimization variable. According to (Mayne et al., 2005; Rawlings and Mayne, 2009), this can be advantageous due to (i) an additional performance improvement by means of the additional degree of freedom and (ii) the fact that the control input $u(t)$ is only a function of the current state $x(t)$ of the real system (2.7) and not depending on the artificial nominal system. Since Ω is an RCI error set, the constraint on the nominal initial state (2.17b) maintains invariance of the tubes. However, this additional degree of freedom dilutes the derivable stability statement allowing only convergence (or asymptotic stability of the composite system) as discussed above. A thorough discussion on this problem can be found in Rawlings and Mayne (2009, Section 3.4).

Remark 2.18. *The nominal state sequence* $\{z^*(0|0), z^*(0|1), z^*(0|2), \dots\}$ *generated by Algorithm 2.16 is not a trajectory of the nominal system (2.12), i.e., in general it follows that* $z^*(0|t+1) \neq f(z^*(0|t), v^*(0|t), 0)$. *However, the nominal initial state can be fixed to follow a trajectory by replacing (2.17b) with* $z(0|t) = z^*(1|t-1) := f(z(0|t-1), v^*(0|t), 0)$ *for all* $t \in \mathbb{I}_{\geq 1}$ *and* $z(0|0) = x_0$ *for* $t = 0$. *Additionally, Algorithm 2.16 must be modified such that it keeps track of the nominal system state. While it can be advantageous in the analysis of the algorithm to consider actual trajectories of the nominal system (2.12), the modified problem is independent of the real state* $x(t)$. *Hence, additional information included in the real system state, for example about past disturbances, is not taken into consideration.*

Computing RCI error sets for general nonlinear functions is impossible. However, approaches have been proposed in order to determine RCI error sets for certain classes of systems. An approach based on incremental input-to-state stability of the system dynamics is proposed in Bayer et al. (2013) (see Yu et al. (2013) for a related continuous-time setup). Similar tubes are derived in (Mayne et al., 2011; Rawlings and Mayne, 2009, Section 3.6), however, they are determined by a different, two-layer setup: While a first MPC determines the nominal trajectory, a second MPC tracks this nominal reference. In Raković et al. (2006), tubes for nonlinear systems with matched nonlinearities are proposed.

Since finding an RCI error set can be cumbersome, other methods not depending on RCI error sets have been presented. All of these methods have some form of tightening in common in order to guarantee recursive feasibility. Based on Lipschitz constants of the dynamics, all possible evolutions of the real system are over-bounded in (Limón et al., 2002a,b; Pin et al., 2009). These predictions can be understood as time-varying tubes, where the variation depends on the Lipschitz constant. Moreover, the authors present ISS for their setup. In special cases, these approaches can also be used in order to determine RCI sets. In Limón et al. (2005), interval arithmetic is used in order to determine the possible evolutions of the real system along the prediction.

---

[6]We revisit this problem in Chapter 4 and provide a possibility to establish asymptotic stability of Ω for the separate closed-loop system (2.19).

Linear systems

A special class of systems thoroughly investigated in robust stabilizing MPC is the class of linear time-invariant systems

$$x(t+1) = Ax(t) + Bu(t) + w(t), \qquad x(0) = x_0, \qquad (2.20)$$

where $A \in \mathbb{R}^{n \times n}$ and $B \in \mathbb{R}^{n \times m}$.

Assumption 2.10. *The linear system (A, B) is stabilizable.*

Considering disturbances $w(t) \in \mathbb{W} \subset \mathbb{R}^n$ satisfying Assumption 2.7, the popularity of this setup lies in the linearity, i.a., allowing the explicit computation of the error evolution independent of the state. When Assumption 2.10 is satisfied, a common choice for the input parametrization is

$$u(t) = v(t) + K\big(x(t) - z(t)\big), \qquad (2.21)$$

where $K \in \mathbb{R}^{m \times n}$ is chosen such that the closed-loop matrix $A_{\mathrm{cl}} := A + BK$ is asymptotically stable, i.e., all eigenvalues are inside the unit disc. With this input parametrization, the error dynamics are

$$e(t+1) = A_{\mathrm{cl}}e(t) + w(t), \qquad e(0) = x_0 - z_0. \qquad (2.22)$$

Using this dynamics and Assumption 2.7, the error at time t satisfies $e(t) \in \Omega_t$, where

$$\Omega_{t+1} = A_{\mathrm{cl}}\Omega_t \oplus \mathbb{W}, \qquad (2.23)$$

with $\Omega_0 = \{x_0 - z_0\}$. Solving this recursion, it follows

$$\Omega_t = A_{\mathrm{cl}}^t \{x_0 - z_0\} \oplus \bigoplus_{k=0}^{t-1} A_{\mathrm{cl}}^k \mathbb{W}. \qquad (2.24)$$

Let us consider the limit set of (2.24) as $t \to \infty$ with $e(0) = 0$. Under Assumption 2.7 and if A_{cl} is asymptotically stable, it was shown that the limit $\Omega_\infty \subset \mathbb{R}^n$ exists and is a compact set (Kolmanovsky and Gilbert, 1998). Moreover, Ω_∞ is not only robust positively invariant with respect to the error dynamics (2.22), but also the minimal robust positively invariant (mRPI) error set. Computing this mRPI set is in general intractable, except for the case that there exists some $k \in \mathbb{I}_{\geq 1}$ and some $\alpha \in [0, 1)$ such that $A_{\mathrm{cl}}^k = \alpha I_n$. Furthermore, several techniques have been introduced to find an invariant approximation of the mRPI set, e.g., (Kouramas et al., 2005; Ong and Gilbert, 2006; Raković et al., 2005).

These mRPI error sets are used for tube-based robust stabilizing MPC of linear systems. In (Mayne et al., 2005; Rawlings and Mayne, 2009), a linear counterpart to Algorithm 2.16 is presented providing similar results to Theorem 2.17. The special case of fixing the nominal initial state—as discussed in Remark 2.18—is considered in (Langson et al., 2004; Rawlings and Mayne, 2009).

In order to reduce the conservatism by considering "infinitely" many disturbances when taking Ω_∞ into consideration, the error tubes (2.24) are considered in Chisci et al. (2001). This requires a few modifications of the optimization problem in order to guarantee recursive feasibility and convergence. However, the tightening of pointwise-in-time constraints is carried out with respect to the less conservative tubes (2.24). Using the input parametrization (2.14), the constraint tightening becomes variant with respect to prediction-time k,

$$\overline{\mathbb{Z}}_k := \mathbb{Z} \ominus (\Omega_k \times K\Omega_k). \qquad (2.25)$$

Assumption 2.11. *The sets $\overline{\mathbb{Z}}_k$ are non-empty and compact for all $k \in \mathbb{I}_{\geq 0} \cup \{\infty\}$.*

In order to be able to apply this tightening, the nominal initial state must be fixed to the measured state at each iteration, i.e., $z(0|t) = x(t)$ leading to $e(0) = 0$. Additionally, the terminal region must be chosen appropriately. This will be discussed in further detail in Chapter 3 and 5.

Due to the linearity of the dynamics, further control schemes are provided in tube-based robust stabilizing MPC for linear systems, which cannot be transferred to nonlinear systems directly. One approach is homothety, meaning the mRPI set or any other set is scaled at each prediction step by a scalar parameter which is an additional optimization variable (Langson et al., 2004; Raković et al., 2012c; Raković et al., 2013). Another approach is a separable prediction scheme, where the input is parametrized by means of all possible future errors bounded in a compact and convex set containing the origin (Löfberg, 2003; Muñoz-Carpintero et al., 2016; Raković et al., 2012a,b).

2.2.3 Stochastic MPC

In contrast to the robust approaches presented in the previous sections, in many practical applications some realizations of the disturbance are more likely than others. However, this is neither considered in min-max robust MPC nor in tube-based robust MPC. In stochastic MPC, this can be represented by equipping the disturbances with probabilistic descriptions providing additional information about the distribution of the disturbance, e.g., by means of probability density functions (PDFs). These can be determined by the underlying physical principles or by stochastic analysis during model identification. This additional information can be utilized to optimize average performance as well as risk measures (Chatterjee and Lygeros, 2015; Muñoz de la Peña et al., 2005; Schwarm and Nikolaou, 1999). By the stochastic information, one can also allow for a certain constraint violation which increases the region of attraction and which can be a reasonable constraint formulation for many practical applications (Cannon et al., 2009a; Primbs and Sung, 2009). On the downside, computational complexity grows rapidly for stochastic MPC approaches such that useful approximation techniques must be found (Blackmore et al., 2010; Bungartz and Griebel, 2004; Deori et al., 2015; Mesbah et al., 2014; Pagnoncelli et al., 2009; Zhang et al., 2014). Moreover, ensuring recursive feasibility is often a difficult aspect (Korda et al., 2011; Kouvaritakis et al., 2010; Lorenzen et al., 2015).

In some setups of stochastic MPC unbounded disturbances are assumed, e.g., Gaussian noise (Bernardini and Bemporad, 2012; Hessem et al., 2001; Ono and Williams, 2008). However, there are also setups where bounded support of the distribution is assumed (Cannon et al., 2011; Kouvaritakis and Cannon, 2015).

We focus on linear time-invariant systems of the form (2.20) with the following assumption on the disturbance and its distribution:

Assumption 2.12. *For each $t \in \mathbb{I}_{\geq 0}$, the disturbance $w(t)$ is modeled by a zero mean random variable with given probability density function (PDF) $\rho_{\mathbb{W}} : \mathbb{R}^n \to [0, \infty)$, which has bounded support $\mathbb{W} \subset \mathbb{R}^n$, where $\mathbb{W}$ is a compact and convex set containing the origin in its interior. Moreover, the sequence of random variables $\{w(t)\}_{t \in \mathbb{I}_{\geq 0}}$ is independent and identically distributed (i.i.d.).*[7]

---

[7]Similar to the discussion of stochastic MPC in Kouvaritakis and Cannon (2015, Section 6.1), we assume

Since we focus on strict satisfaction of the constraints and only use the additional information in order to improve the performance, chance constraints, that is, constraints that need only be satisfied with given probability, are not considered in the following discussion.

With a distribution provided, it is possible to optimize over the expected cost

$$J_N(x(t), \boldsymbol{u}(t)) = \sum_{k=0}^{N-1} \mathbb{E}\left\{\ell\Big(x(k|t), u(k|t)\Big)\Big| x(t)\right\} + \mathbb{E}\left\{V_{\mathrm{f}}\Big(x(N|t)\Big)\Big| x(t)\right\} \quad (2.26)$$

instead of considering some nominal or some worst-case objective as in (2.18) or (2.11), respectively. Here again, ℓ is the stage cost function and V_{f} is the terminal cost function. The only information available at time t is the current state $x(t)$. Hence, the expectation is conditional with respect to this information and depends on the distribution of the unknown future disturbance sequence $\boldsymbol{w}(t) = \{w(t), \dots, w(t+N-1)\}$.

For the special case of a quadratic stage cost function

$$\ell(x, u) = x^\top Q x + u^\top R u, \quad (2.27)$$

with $Q \succ 0$ and $R \succ 0$, results on the average performance have been presented in the literature. Again, the input parametrization (2.14) is used with K chosen such that A_{cl} is asymptotically stable. We employ the terminal cost function

$$V_{\mathrm{f}}(x) = x^\top P x, \quad (2.28)$$

where P is the solution to the discrete-time Lyapunov equation $P = A_{\mathrm{cl}}^\top P A_{\mathrm{cl}} + Q + K^\top R K$. Moreover, the terminal region is given by $\overline{\mathbb{X}}_{\mathrm{f}} = O_{\max} \ominus \Omega_N$, where

$$O_{\max} := \left\{ x \in \mathbb{X} : \left(A_{\mathrm{cl}}^k x, K A_{\mathrm{cl}}^k x\right) \in \overline{\mathbb{Z}}_k, \forall k \in \mathbb{I}_{\geq 0} \right\} \quad (2.29)$$

is the maximal output admissible set on $\mathbb{Z}$ with respect to the origin and the disturbance $\mathbb{W}$ (Gilbert and Tan, 1991; Kolmanovsky and Gilbert, 1995, 1998). Together with the prediction horizon $N \in \mathbb{I}_{\geq 1}$, we can state the following optimization problem to determine the receding horizon controller at time t with $x(t)$ given:

Problem 2.19.

$$\underset{\boldsymbol{v}(t)}{\text{minimize}} \; J_N\Big(x(t), \boldsymbol{v}(t)\Big)$$

subject to

$$z(k+1|t) = Az(k|t) + Bv(k|t), \qquad \forall k \in \mathbb{I}_{[0,N-1]}, \quad (2.30a)$$
$$z(0|t) = x(t), \quad (2.30b)$$
$$\Big(z(k|t), v(k|t)\Big) \in \overline{\mathbb{Z}}_k, \qquad \forall k \in \mathbb{I}_{[0,N-1]}, \quad (2.30c)$$
$$z(N|t) \in \overline{\mathbb{X}}_{\mathrm{f}}, \quad (2.30d)$$

where $J_N(x(t), \boldsymbol{v}(t))$ is given in (2.26) with ℓ from (2.27) and V_{f} from (2.28).

---

the disturbance w to be a random variable whose variable $w(t)$ at time t is unknown but has a known probability distribution. Hence, every random event is related straightforward to the realizations of $w(t)$. With this understanding, the underlying probability space is well defined and measure-theoretic notation can be avoided.

We denote the set of states $x \in \mathbb{X}$ for which there exists a solution to Problem 2.19 with $x(t) = x$ by $\tilde{\mathbb{X}}_N := \{x \in \mathbb{R}^n : \exists \boldsymbol{v} = \{v(0), \ldots, v(N-1)\} \text{ s.t. } z(0) = x, z(k+1) = f(z(k), v(k), 0), (z(k), v(k)) \in \overline{\mathbb{Z}}_k, \forall k \in \mathbb{I}_{[0,N-1]}, \text{ and } z(N) \in \overline{\mathbb{X}}_f\}$.

The optimization problem is applied in the following algorithm:

Algorithm 2.20 (Robust model predictive control with stochastic information). *At each time instant $t \in \mathbb{I}_{\geq 0}$, measure the state $x(t)$ and solve Problem 2.19. Apply the control input $u(t) := v^*(0|t)$ to system (2.20).*

Considering the closed-loop system resulting from applying Algorithm 2.20, the following result on average performance and (recursive) feasibility is provided:

Theorem 2.21. *Let $x_0 \in \tilde{\mathbb{X}}_N$. Suppose that Assumptions 2.10–2.12 are satisfied. Then, Problem 2.19 is feasible for all $t \in \mathbb{I}_{\geq 0}$. Furthermore, the closed-loop system resulting from applying Algorithm 2.20 satisfies the pointwise-in-time constraints (2.2) for all $t \in \mathbb{I}_{\geq 0}$ and its average performance satisfies*

$$\lim_{T \to \infty} \frac{1}{T} \sum_{t=0}^{T-1} \mathbb{E}\left\{ x(t)^\top Q x(t) \,\middle|\, x_0 \right\} \leq \mathbb{E}\left\{ w^\top P w \right\}. \tag{2.31}$$

We note that the average performance (2.31) is provided as a conditional expectation with respect to the information available at time 0, that is, it is conditional with respect to x_0.

The proof for this result follows along the lines of the results in (Cannon et al., 2009b; Kouvaritakis and Cannon, 2015; Lorenzen et al., 2015) and can be seen as a special case of these results. While all of the aforementioned references take chance constraints into account, here all constraints are satisfied with probability 1 since a strict tightening is considered in (2.30c) by means of the idea in Chisci et al. (2001).

2.3 Economic MPC

Instead of stabilizing an a priori determined steady-state, *economic MPC* aims at optimizing the performance of a system with respect to some arbitrary stage cost functions. While in stabilizing MPC, the objective is usually chosen to be positive definite with respect to the predetermined behavior, see, e.g., Assumption 2.5, *no* such condition is imposed in economic MPC, which means Assumption 2.5 no longer holds. This allows for more general stage cost functions being directly associated to the economics of the underlying process. However, this can lead to cyclic or even chaotic behavior, hence, the optimal closed-loop behavior is not as obvious as in the stabilizing case and we might not achieve convergence to an equilibrium with the closed loop.

The goal in economic MPC is to find a feasible control input to system (2.1) minimizing the asymptotic average cost

$$\limsup_{T \to \infty} \frac{1}{T} \sum_{t=0}^{T-1} \ell\big(x(t), u(t)\big), \tag{2.32}$$

and the main theoretical idea aims at providing bounds on (2.32). In economic MPC, the performance is usually compared to the best steady-state behavior (see, e.g., Angeli et al. (2012); Ellis et al. (2014); Rawlings et al. (2012)), but there are also results on periodic solutions (see, e.g., Limón et al. (2014); Müller and Grüne (2016); Zanon et al. (2013)). Following the first approach, we introduce the *optimal steady-state* as the feasible equilibrium minimizing the stage cost

$$(x_s, u_s) = \arg\min_{x=f(x,u),\,(x,u)\in\mathbb{Z}} \ell(x, u), \tag{2.33}$$

which exists under Assumptions 2.1–2.3. There need not exist a unique optimal steady-state. In these cases, one can be chosen. Regarding the optimal steady-state, we can derive the following statement on the performance bound, given in (Amrit et al., 2011; Angeli et al., 2012; Müller, 2014), where the algorithm and the assumptions are understood with respect to an arbitrary stage cost function ℓ, *not* necessarily satisfying Assumption 2.5:

Theorem 2.22. *Let $x_0 \in \mathbb{X}_N$. Suppose that Assumptions 2.1–2.4 and 2.6 are satisfied and let (x_s, u_s) be the optimal steady-state according to (2.33). Then, Problem 2.2 is feasible for all $t \in \mathbb{I}_{\geq 0}$. Furthermore, the closed-loop system (2.6) resulting from applying Algorithm 2.3 satisfies the pointwise-in-time constraints (2.2) for all $t \in \mathbb{I}_{\geq 0}$ and it has an asymptotic average performance which is no worse than that of the optimal steady-state, i.e.,*

$$\limsup_{T\to\infty} \frac{1}{T} \sum_{t=0}^{T-1} \ell\big(x(t), u^*(0|t)\big) \leq \ell(x_s, u_s). \tag{2.34}$$

We take a look at the main idea of the proof as it becomes relevant in the subsequent sections. Using standard ideas, one can show that

$$0 \leq \liminf_{T\to\infty} \frac{1}{T} \Big(V_N\big(x(T)\big) - V_N\big(x(0)\big)\Big) = \liminf_{T\to\infty} \frac{1}{T} \sum_{t=0}^{T-1} \Big(V_N\big(x(t+1)\big) - V_N\big(x(t)\big)\Big)$$

$$\leq \ell(x_s, u_s) - \limsup_{T\to\infty} \frac{1}{T} \sum_{t=0}^{T-1} \ell\big(x(t), u^*(0|t)\big).$$

The first inequality follows from continuity of ℓ and V_f and the compactness of $\mathbb{Z}$; the second inequality follows from Assumption 2.6(iii) leading to $V_N(x(t+1)) - V_N(x(t)) \leq -\ell(x(t), u^*(0|t)) + \ell(x_s, u_s)$. We note that in Angeli et al. (2012) a terminal equality constraint is considered, i.e., $\mathbb{X}_f = x_s$. In Amrit et al. (2011), a constructive approach is provided to compute V_f, $\mathbb{X}_f$, and κ_f satisfying Assumption 2.6 for arbitrary $(\mathcal{C}^2)$ stage cost functions.

Remark 2.23. *Considering Assumption 2.6, we required the optimal steady-state x_s to be in the* interior *of the terminal region $\mathbb{X}_f$. In case of economic MPC, this is only necessary if stability of the economic MPC approach is investigated (see, e.g., Theorem 2.28 and Amrit et al. (2011)). If one is interested in the asymptotic average performance or optimal operating regimes (see subsequent section), Assumption 2.6 can be relaxed to $x_s \in \mathbb{X}_f$. The same arguments apply in the following sections as well, however, for ease of presentation, we always state the respective assumptions on the terminal region with $x_s \in \mathrm{int}(\mathbb{X}_f)$.*

2.3.1 Optimal operation and stability in economic MPC

As mentioned before and in contrast to stabilizing MPC, it is not necessarily optimal to drive the system to the optimal steady-state. Thus, the desired or expected closed-loop behavior need not be known, and may especially not be related to some steady-state. In fact, the optimal operating regime is not obvious. Even though we are not interested in tracking an a priori defined behavior, it can still be interesting to investigate if the optimal behavior has a certain structure, especially since economic MPC does not rule out optimal operation at a steady-state. In the nominal case, this has for example been investigated in (Angeli et al., 2012; Müller et al., 2015a,b) for steady-state optimality and in (Grüne and Zanon, 2014; Müller and Grüne, 2016; Zanon et al., 2016b) for optimality of a periodic orbit.

For steady-state optimality, we investigate whether steady-state operation is optimal for a problem setup in economic MPC in the following sense:

Definition 2.24 (Optimal operation at steady-state). *System* (2.1) *is* optimally operated at steady-state *with respect to the cost function ℓ, if for each solution satisfying $(x(k), u(k)) \in \mathbb{Z}$ for all $k \in \mathbb{I}_{\geq 0}$ it holds that*

$$\liminf_{T \to \infty} \frac{1}{T} \sum_{t=0}^{T-1} \ell\big(x(t), u(t)\big) \geq \ell(x_s, u_s), \tag{2.35}$$

where (x_s, u_s) is the optimal steady-state as defined in (2.33).

This definition means that there is no feasible closed-loop sequence providing a better asymptotic average performance than the steady-state.

In the literature on economic MPC, dissipativity turns out to be relevant for proving if a system is optimally operated at steady-state (see, e.g., Angeli and Rawlings (2010); Angeli et al. (2012)). Dissipativity was introduced in Willems (1972) and adapted to discrete-time systems in Byrnes and Lin (1994). We use the following adopted notion introduced in (Angeli et al., 2012; Müller et al., 2015a), where for a given set $\mathbb{S} \subseteq \mathbb{Z}$ the projection onto $\mathbb{X}$ is denoted by $\mathbb{S}_{\mathbb{X}}$, i.e., $\mathbb{S}_{\mathbb{X}} := \{x \in \mathbb{X} : \exists u \in \mathbb{U} \text{ s.t. } (x, u) \in \mathbb{S}\}$:

Definition 2.25 (Dissipativity). *System* (2.1) *is* dissipative *on a set $\mathbb{S} \subseteq \mathbb{Z}$ with respect to the supply rate $s : \mathbb{X} \times \mathbb{U} \to \mathbb{R}$ if there exists a storage function $\lambda : \mathbb{S}_{\mathbb{X}} \to \mathbb{R}_{\geq 0}$ such that the following inequality is satisfied for all $(x, u) \in \mathbb{S}$:*

$$\lambda\big(f(x, u)\big) - \lambda(x) \leq s(x, u). \tag{2.36}$$

If, in addition, for some $\rho \in \mathcal{K}_{\infty}$ it holds that for all $(x, u) \in \mathbb{S}$

$$\lambda\big(f(x, u)\big) - \lambda(x) \leq -\rho\big(|x - x_s|\big) + s(x, u), \tag{2.37}$$

system (2.1) *is* strictly dissipative *on $\mathbb{S}$.*

In order to limit the investigation to feasible trajectories, we introduce

$$\mathbb{Z}^o := \Big\{(x, u) \in \mathbb{Z} : \exists \boldsymbol{v} \text{ s.t. } \big(z(0), v(0)\big) = (x, u), z(k+1) = f\big(z(k), v(k)\big),$$
$$\big(z(k), v(k)\big) \in \mathbb{Z} \; \forall k \in \mathbb{I}_{\geq 0}\Big\}, \tag{2.38}$$

and the associated set $\mathbb{X}^o$ is the projection of $\mathbb{Z}^o$ onto $\mathbb{X}$, i.e, $\mathbb{X}^o := \{x \in \mathbb{X} : \exists u \in \mathbb{U} \text{ s.t. } (x, u) \in \mathbb{Z}^o\}$. The set $\mathbb{Z}^o$ is the set of all state and input pairs such that there exists a trajectory feasible for all subsequent time instances. With these definitions at hand, we can recall the result on sufficiency of dissipativity for optimal operation at steady-state presented in Angeli et al. (2012):

Theorem 2.26. *Suppose that system* (2.1) *is dissipative on* $\mathbb{Z}^o$ *with respect to the supply rate*

$$s(x, u) = \ell(x, u) - \ell(x_s, u_s), \tag{2.39}$$

then the system is optimally operated at steady-state.

Recently, investigations considering the necessity of dissipativity for optimal operation at steady-state have been conducted (see, e.g., Müller et al. (2013); Müller et al. (2015a)). The results are based on a specific controllability assumption provided by means of the following sets. Given the optimal steady-state x_s and $M \in \mathbb{I}_{\geq 1}$, $\mathcal{C}_M$ denotes the controllable set, i.e., the set of all states that can be steered to x_s in M steps,

$$\mathcal{C}_M := \Big\{ x \in \mathbb{X} : \exists \, \boldsymbol{v} \text{ s.t. } z(0) = x, z(k+1) = f\big(z(k), v(k)\big), z(M) = x_s,$$
$$\big(z(k), v(k)\big) \in \mathbb{Z} \; \forall k \in \mathbb{I}_{[0,M-1]} \Big\}, \tag{2.40}$$

and $\mathcal{R}_M$ the set of all states reachable from x_s in M steps

$$\mathcal{R}_M := \Big\{ x \in \mathbb{X} : \exists \, \boldsymbol{v} \text{ s.t. } z(0) = x_s, z(k+1) = f\big(z(k), v(k)\big), z(M) = x,$$
$$\big(z(k), v(k)\big) \in \mathbb{Z} \; \forall k \in \mathbb{I}_{[0,M-1]} \Big\}. \tag{2.41}$$

Furthermore, the set of feasible state/input pairs which are part of a sequence pair $(\boldsymbol{z}, \boldsymbol{v})$ such that $z(\cdot) \in \mathcal{C}_M \cap \mathcal{R}_M$ for all times is

$$\mathcal{Z}_M := \Big\{ (x, u) \in \mathbb{Z} : \exists \, \boldsymbol{v} \text{ s.t. } \big(z(0), v(0)\big) = (x, u), z(k+1) = f\big(z(k), v(k)\big),$$
$$\big(z(k), v(k)\big) \in \mathbb{Z}, z(k) \in \mathcal{C}_M \cap \mathcal{R}_M \; \forall k \in \mathbb{I}_{[0,M-1]} \Big\} \subseteq \mathbb{Z}^o. \tag{2.42}$$

The projection of $\mathcal{Z}_M$ onto $\mathbb{X}$ is given by $\mathcal{X}_M := \{ x \in \mathbb{X} : \exists u \in \mathbb{U} \text{ s.t. } (x, u) \in \mathcal{Z}_M \}$.

With the controllability assumption inherently given in $\mathcal{Z}_M$, necessity of dissipativity for optimal steady-state operation can be provided:

Theorem 2.27. *Suppose that system* (2.1) *is optimally operated at steady-state and let Assumptions 2.1–2.3 be satisfied. Then, for each* $M \in \mathbb{I}_{\geq 1}$, *system* (2.1) *is dissipative on* $\mathcal{Z}_M$ *with respect to the supply rate* (2.39).

The proof of this result is based on the relation between the available storage and dissipativity (see, e.g., Willems (1972)). One tries to find a trajectory which is non-dissipative but still satisfies the optimality condition in (2.35), which leads to a contradiction. In Müller et al. (2015a), it is furthermore shown that a local controllability assumption (linearization of the system is controllable at (x_s, u_s)) is sufficient for dissipativity if the system satisfies the stricter condition of *uniform suboptimal operation off steady-state*.[8]

By the converse result (necessity of dissipativity under the controllability assumption), one can see that dissipativity is not a too strict requirement for a system to satisfy in

---

[8]A system is *suboptimally operated off steady-state* if it is optimally operated at steady-state and satisfies at least one of the two following conditions: (1) Statement (2.35) holds with strict inequality; (2) $\liminf_{k \to \infty} |x(k) - x_s| = 0$. Considering the converse direction, it was shown in Angeli et al. (2012) that a system is suboptimally operated off steady-state if the system is *strictly dissipative*.

order to provide optimal operation at steady-state, even though there might exist systems that are optimally operated at steady-state but not dissipative (compare to Müller (2014); Müller et al. (2015a)).

Considering stability, it turns out that (strict) dissipativity is essential as well. A first stability result in the economic framework was established in Diehl et al. (2011) (limited to linear storage functions) and extended in (Amrit et al., 2011; Angeli et al., 2012).

Theorem 2.28. *Suppose that Assumptions 2.1–2.4 and 2.6 are satisfied, that system* (2.1) *is strictly dissipative on* $\mathbb{Z}^o$ *with respect to the supply rate* (2.39), *and that the storage function* $\lambda(\cdot)$ *is continuous at* x_s. *Then,* x_s *is an asymptotically stable equilibrium of the closed-loop system* (2.6) *with region of attraction* $\mathbb{X}_N$.

The proof was first established in Diehl et al. (2011) (under a strong duality instead of a strict dissipativity assumption) and in Angeli et al. (2012), both with a terminal equality constraint, i.e., $\mathbb{X}_f = \{x_s\}$. We note that both approaches need an additional weak controllability assumption. The theorem above is established in Amrit et al. (2011) under the assumption that the storage function is continuous on the whole domain $\mathbb{Z}$. In Müller (2014), it is argued that continuity at x_s is enough in order to establish stability in the sense of Definition 2.1. Similar statements can also be found in the case of economic MPC without terminal constraints (see, e.g., Damm et al. (2014); Grüne (2013); Grüne and Stieler (2014)), and when combining economic and stabilizing approaches (Maree and Imsland, 2016; Zavala, 2015). Another idea is presented in Zanon et al. (2016a), where local equivalence of economic and tracking MPC is investigated in order to use the latter to show stability of the former. Periodic stability of economic MPC is investigated in Zanon et al. (2016b).

Remark 2.29. *In all of the proofs mentioned above, existence of an appropriate storage function is required. However, no storage function must be provided within the algorithm. The same holds true for the sets required in the necessity result. For some special classes of systems, for example linear dynamics with strictly convex cost functions, storage functions can be determined (see, e.g., Damm et al. (2014); Diehl et al. (2011)). However, there is no general procedure to determine an appropriate storage function.*

2.4 Summary

In this chapter, we presented an overview over the basic results in both nominal and robust stabilizing MPC as well as in economic MPC. We recapped that for stabilizing MPC, i.e., when the stage cost function is positive definite with respect to some desired setpoint (or set), this setpoint (set) is asymptotically stable under suitable assumptions for the closed-loop system under application of the MPC algorithm. This holds for both undisturbed and disturbed systems. In economic MPC, i.e., in the case that no definiteness assumptions are imposed on the stage cost function, we derived bounds on the asymptotic average performance of the closed-loop system and showed that convergence to an equilibrium need not be optimal. However, under suitable dissipativity conditions optimal steady-state operation as well as stability were shown.

Chapter 3

Min-max robust economic MPC

In this chapter, we present a first concept to consider information about disturbances within an economic MPC framework. The basic idea is to incorporate the disturbances by means of optimizing the input for the worst case along the predictions. Within this section, we investigate two different approaches: the first approach (Section 3.1) is inspired by the classical min-max stabilizing MPC presented in Section 2.2.1, the second approach (Section 3.2) combines the min-max idea with concepts from tube-based MPC in order to reduce the computational complexity.

The results presented in this chapter are based on Bayer et al. (2016b).

3.1 Economic MPC in classical min-max setup

As discussed in Section 2.3, our main objective is to find a feasible control input for the nonlinear disturbance-affected system

$$x(t+1) = f\Big(x(t), u(t), w(t)\Big), \qquad x(0) = x_0, \tag{3.1}$$

such that it minimizes the asymptotic average cost

$$\limsup_{T \to \infty} \frac{1}{T} \sum_{t=0}^{T-1} \ell\Big(x(t), u(t)\Big), \tag{3.2}$$

where ℓ is again some arbitrary—economic—continuous objective. Similar to the nominal economic MPC approach, we derive an MPC scheme which provides bounds on the asymptotic average performance (3.2) under the disturbance $w(t)$ satisfying Assumption 2.7.

From a conceptual point of view, Problem 2.10 provides the most general min-max setup since it allows for optimizing over general nonlinear feedback laws. Even though Problem 2.10 is provided for min-max robust stabilizing MPC, the terminal ingredients can be modified in order to be applicable for min-max robust *economic* MPC. However, as discussed in Section 2.2.1, this problem would be computationally intractable. Thus, we introduce a scheme following the approaches in (Limón et al., 2009; Raimondo et al., 2009). Due to the structural similarity of these approaches with the setup considered in this section, we refer to the new min-max robust economic MPC scheme introduced in this section as the *classical* setup. In Limón et al. (2009), computational feasibility is regained by a given continuous input parametrization (compare to (2.14))

$$u(t) = \pi\Big(x(t), c(t)\Big), \tag{3.3}$$

where $\pi : \mathbb{R}^n \times \mathbb{R}^m \to \mathbb{R}$ and $c(t) \in \mathbb{R}^m$ is the manipulated input. This parametrization is in between open-loop and feedback control (sometimes referred to as *semi-feedback* formulation) since we allow for a (fixed) feedback law while using c as the optimization variable. In contrast to the parametrization in Section 2.2.2, the one in (3.3) is independent of any nominal dynamics. However, if a nominal system is to be considered (as we discuss subsequently), the same parametrization must be employed for the nominal input.

Using the parametrization, we can simplify the notation introducing

$$f_\pi\big(x(t), c(t), w(t)\big) := f\big(x(t), \pi(x(t), c(t)), w(t)\big) \tag{3.4}$$

$$\text{and} \qquad \ell_\pi\big(x(t), c(t)\big) := \ell\big(x(t), \pi(x(t), c(t))\big). \tag{3.5}$$

In order to take all possible open-loop states into account, we introduce state tubes with initial condition $x(0) = x_0$ defined by

$$X_{t+1} = \Phi_\pi\big(X_t, c(t), \mathbb{W}\big), \qquad X_0 = \{x_0\}, \tag{3.6}$$

where $\Phi_\pi : 2^{\mathbb{R}^n} \times \mathbb{R}^m \times 2^{\mathbb{R}^q} \to 2^{\mathbb{R}^n}$ describes the set of all possible states under the dynamics f and the parametrized input π, i.e.,

$$\Phi_\pi\big(X_t, c(t), \mathbb{W}\big) := \big\{ f_\pi\big(x, c(t), w\big) : x \in X_t, w \in \mathbb{W} \big\}. \tag{3.7}$$

Computing these tubes for nonlinear dynamics is in general a difficult task. To this end, some approximation-based results have been introduced in the literature (see Limón et al. (2009) for a detailed discussion on this topic). For linear dynamics with a linear input parametrization of the form

$$\pi(x, c) = Kx + c, \tag{3.8}$$

with $K \in \mathbb{R}^{m \times n}$, computing the tubes is simplified significantly, and the tubes can be computed exactly (see, e.g., Chisci et al. (2001); Scokaert and Mayne (1998)).

With these state tubes at hand, we can compute the receding horizon control law at each time $t \in \mathbb{I}_{\geq 0}$ for a given $x(t)$:

Problem 3.1.
$$\underset{c(t)}{\text{minimize}} \ \underset{w(t)}{\text{maximize}} \ J_N^c\big(x(t), \boldsymbol{c}(t), \boldsymbol{w}(t)\big)$$

subject to

$$x(k+1|t) = f_\pi\big(x(k|t), c(k|t), w(k|t)\big), \qquad \forall k \in \mathbb{I}_{[0,N-1]}, \tag{3.9a}$$

$$x(0|t) = x(t), \tag{3.9b}$$

$$w(k|t) \in \mathbb{W}, \qquad \forall k \in \mathbb{I}_{[0,N-1]}, \tag{3.9c}$$

$$X_{k+1} = \Phi_\pi(X_k, c(k|t), \mathbb{W}), \qquad \forall k \in \mathbb{I}_{[0,N-1]}, \tag{3.9d}$$

$$X_0 = \{x(t)\}, \tag{3.9e}$$

$$\big(x_k, \pi(x_k, c(k|t))\big) \in \mathbb{Z}, \qquad \forall x_k \in X_k, \forall k \in \mathbb{I}_{[0,N-1]}, \tag{3.9f}$$

$$X_N \subseteq \mathbb{X}_\mathrm{f}, \tag{3.9g}$$

where

$$J_N^c\big(x(t), \boldsymbol{c}(t), \boldsymbol{w}(t)\big) = \sum_{k=0}^{N-1} \ell_\pi\big(x(k|t), c(k|t)\big) + V_\mathrm{f}^c\big(x(N|t)\big). \tag{3.10}$$

We denote the predicted disturbance sequence and the predicted input sequence by $\boldsymbol{w}(t) = \{w(0|t), \ldots, w(N-1|t)\}$ and $\boldsymbol{c}(t) = \{c(0|t), \ldots, c(N-1|t)\}$, respectively. The value function of this problem is denoted by $V_N^c(x(t))$ and the arguments of the optimum by $\boldsymbol{c}^*(t)$ and $\boldsymbol{w}^*(t)$.

Before introducing and discussing the terminal ingredients of Problem 3.1, we introduce the optimal steady-state (x_s^c, c_s^c) satisfying

$$(x_s^c, c_s^c) = \underset{x=f_\pi(x,c,0),\,(x,c)\in\overline{\mathcal{Z}}}{\arg\min} \ell_\pi(x, c), \tag{3.11}$$

where the constraints are given by

$$\overline{\mathcal{Z}} := \big\{(x,c) \in \mathbb{R}^n \times \mathbb{R}^m : \bar{x}(0) = x, \bar{x}(t+1) = f_\pi(\bar{x}(t), c, w(t)),$$
$$\big(\bar{x}(t), \pi(\bar{x}(t), c)\big) \in \mathbb{Z}, \forall w(t) \in \mathbb{W} \text{ and } \forall t \in \mathbb{I}_{\geq 0}\big\}. \tag{3.12}$$

Existence of the optimal steady-state follows by Assumptions 2.1–2.3 and continuity of the input parametrization. Besides considering the parametrized input, the above definition of the optimal steady-state differs from (2.33) by the constraints. We only consider those state and input pairs that guarantee feasibility of the pointwise-in-time constraint (2.2) for all possible disturbances and for all times when applying the constant input c at each time instance. For the optimal steady-state this means in particular that the real system state will always be feasible when starting at x_s^c and applying c_s^c. Later, a reasonable approximation of the set of steady-states in $\overline{\mathcal{Z}}$ is provided by means of tightening $\mathbb{Z}$ appropriately.

Considering the terminal cost, the terminal region as well as the terminal controller, we state the following assumption:

Assumption 3.1. *There exists a terminal region $\mathbb{X}_f \subseteq \mathbb{X}$, containing the optimal steady-state x_s^c, a terminal cost $V_f^c : \mathbb{X}_f \to \mathbb{R}$, and a class $\mathcal{K}_\infty$ function ρ such that for all $x \in \mathbb{X}_f$ and all $w \in \mathbb{W}$ it holds that*

(i) $\big(x, \pi(x, c_s^c)\big) \in \mathbb{Z}$,

(ii) $f_\pi(x, c_s^c, w) \in \mathbb{X}_f$,

(iii) $V_f^c\big(f_\pi(x, c_s^c, w)\big) - V_f^c(x) \leq -\ell_\pi\big(x, c_s^c\big) + \rho(|w|) + \ell_\pi(x_s^c, c_s^c)$.

Moreover, the terminal cost V_f^c is continuous on $\mathbb{X}_f$.

As stated in Limón et al. (2009), this requires that $\pi(x, c_s^c)$ asymptotically stabilizes the corresponding steady-state x_s^c in the case that no disturbances are present.

We define the controller in this classical min-max setup by the following algorithm:

Algorithm 3.2 (Classical min-max robust economic model predictive control). *At each time instant $t \in \mathbb{I}_{\geq 0}$, measure the state $x(t)$ and solve Problem 3.1. Apply the control input $u(t) := \pi(x(t), c^*(0|t))$ to system (3.1).*

Denoting the set of all states x for which Problem 3.1 is feasible with prediction horizon N and $x(t) = x$ by $\overline{\mathbb{X}}_N^c := \{x \in \mathbb{R}^n : \exists \boldsymbol{c} = \{c(0), \ldots, c(N-1)\} \text{ s.t. } X_0 = \{x\}, X_{k+1} = \Phi_\pi(X_k, c(k), \mathbb{W}), (x_k, \pi(x_k, c(k))) \in \mathbb{Z}, \forall x_k \in X_k, \forall k \in \mathbb{I}_{[0,N-1]} \text{ and } X_N \subseteq \mathbb{X}_f\}$, one obtains the following result on feasibility and average performance of the closed-loop system which results from applying Algorithm 3.2:

Theorem 3.3. *Let $x_0 \in \overline{\mathbb{X}}_N^c$. Suppose that Assumptions 2.1–2.3, 2.7, and 3.1 are satisfied and let (x_s^c, c_s^c) be the optimal steady-state according to (3.11). Then, Problem 3.1 is feasible for all $t \in \mathbb{I}_{\geq 0}$. Furthermore, the closed-loop system which results from applying Algorithm 3.2 satisfies the pointwise-in-time constraints (2.2) for all $t \in \mathbb{I}_{\geq 0}$ as well as the following bound on the asymptotic average performance:*

$$\limsup_{T \to \infty} \frac{1}{T} \sum_{t=0}^{T-1} \ell_\pi \Big(x(t), c^*(0|t) \Big) \leq \ell_\pi(x_s^c, c_s^c) + \rho(w_{\max}). \tag{3.13}$$

When investigating the performance bound stated in Theorem 3.3, we notice that the bound is on the performance of the *real* closed loop. Because of the special structure of the min-max robust economic MPC problem, the bound holds for all possible realizations of the disturbance sequence (while the disturbance satisfies Assumption 2.7). However, the bound might be conservative (i) as it has to hold for *any* possible realization of the disturbance sequence, and (ii) the bound is provided by comparison functions which usually lead to conservative over-approximations.

The major problem concerning this approach and in particular solving Problem 3.1 is the computational complexity. Even though we parametrize the input using (3.3), solving the min-max problem remains difficult. As discussed in Section 2.2.1, there exist several methods for solving this problem in the literature based on, e.g., dynamic programming or combinatorial approaches. However, for general problems, there can only be approximated solutions.

Another challenge in this setup besides the computational complexity is to find a terminal cost V_f^c along with an associated class $\mathcal{K}_\infty$ function ρ such that Assumption 3.1(iii) is satisfied. While it is often possible to find *some* functions V_f^c and ρ satisfying this assumption, in general it is difficult to find a ρ which provides a small over-approximation.

One possible approach to find these functions that can provide a good bound stems from the literature on input-to-state stability (ISS). Since the structure of Assumption 3.1(iii) is similar to the problem of proving ISS, methods developed in this framework can lead to reasonable V_f^c and ρ. Exemplarily, we mention the approach in Sontag and Wang (1995, Remark 4).

3.2 Economic MPC in min-max setup based on tubes

In this section, we present an approach for min-max robust economic MPC approximating the exact, classical approach introduced in Section 3.1. This new approach is inspired by concepts from tube-based robust MPC schemes in stabilizing MPC (see, e.g., Chisci et al. (2001); Mayne et al. (2005)). The main idea is the following: Instead of maximizing over the disturbance along the whole prediction, we consider the worst case at each prediction step individually, that is, instead of maximizing the sum with respect to the disturbance, we maximize each term within the sum. The resulting problem will be more conservative than the one considered in Problem 3.1, however, the computational complexity can reduce significantly. For some classes of problems, it is even possible to replace the maximization by an analytic expression which further simplifies the optimization.

The key idea in min-max robust MPC based on tubes is to introduce the (artificial) nominal system given by

$$z(t+1) = f_\pi(z(t), c(t), 0), \qquad z(0) = z_0 \tag{3.14}$$

(compare to (2.12)). With this nominal system, we recall the error $e(t) := x(t) - z(t)$ (see (2.13)), with its associated dynamics

$$e(t+1) = f_\pi(x(t), c(t), w(t)) - f_\pi(z(t), c(t), 0), \qquad e(0) = x_0 - z_0. \tag{3.15}$$

We note that due to the input parametrization (3.3), the manipulated input $c(t)$ is applied to both systems, the real and the nominal.

Remark 3.4. *The input parametrizations (2.14) and (3.3) are not equivalent in the sense that there need not exist a pair of φ and π such that the error dynamics (2.15) and (3.15) are equivalent. This holds only for some special cases, e.g., linear systems with linear parametrizations. For continuity of presentation, we remain with the parametrization by π in this chapter.*

The error dynamics can be used to determine bounds on the evolution of the error. Similar to Definition 2.13, one defines the robust control invariant error set taking the new input parametrization (3.3) into account:

Definition 3.5 (Robust control invariance (RCI)—adapted). *Suppose that Assumption 2.7 is satisfied. A set $\Omega \subseteq \mathbb{R}^n$ is robust control invariant (RCI) for the error dynamics (3.15) if there exists a feedback control law $u(t) = \pi(x(t), c(t))$ such that for all $x(t), z(t) \in \mathbb{R}^n$ and $c(t) \in \mathbb{R}^m$ with $e(t) := x(t) - z(t) \in \Omega$ and $\big(x(t), \pi(x(t), c(t))\big) \in \mathbb{Z}$, and for all $w(t) \in \mathbb{W}$, it holds $e(t+1) \in \Omega$.*

Using the definition of the RCI set, it holds that $x(t) \in \{z(t)\} \oplus \Omega$ for all $t \in \mathbb{I}_{\geq 0}$ if $e(0) \in \Omega$. Thus, by means of the nominal system and the RCI set Ω, bounds can be provided on the behavior of the real system state. Also for these adapted RCI sets, we suppose Assumption 2.8 to hold.

The relation of the predictions of the nominal system combined with the RCI set Ω and the tubes in the classical case is depicted in Figure 3.1: The uncertainty affected states can be predicted within the exact tubes X_k as computed in (3.6). One can see that (i) the exact tubes X_k are always within the RCI set centered at the nominal predictions $\{z(k|t)\} \oplus \Omega$ (as long as $x(t) - z(0|t) \in \Omega$), (ii) the initial states of the tubes $x(t)$ and $z(0|t)$ do not need to intersect, and (iii) the nominal dynamics are restricted to the tightened set $\overline{\mathbb{X}}$, being the projection of the tightened constraint set

$$\overline{\mathbb{Z}} := \Big\{ (z, c) \in \mathbb{R}^n \times \mathbb{R}^m : \big(x, \pi(x, c)\big) \in \mathbb{Z} \text{ for all } x \in \{z\} \oplus \Omega \Big\} \tag{3.16}$$

onto the state space $\mathbb{X}$, i.e., $\overline{\mathbb{X}} := \Big\{ z \in \mathbb{X} : \exists c \in \mathbb{R}^m \text{ s.t. } (z, c) \in \overline{\mathbb{Z}} \Big\}$. We note the slight difference to the constraint tightening in the previous section (2.16) due to the different input parametrization.

Determining the RCI sets can be difficult but there are several approaches in the literature (see discussion in Section 2.2.2). In case of linear systems, one can use robust positively invariant (RPI) sets, which are usually more simple to determine.

The receding horizon control law for the min-max economic MPC based on tubes is determined by solving the following optimization problem at each time $t \in \mathbb{I}_{\geq 0}$ for a given $x(t)$:

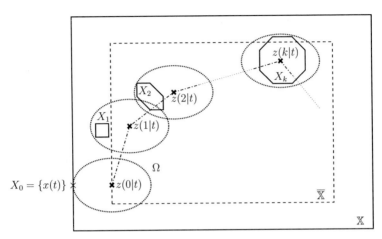

Figure 3.1: Schematic relation of the exact tubes X_k (solid) derived according to (3.6) and the RCI set Ω (dotted). The nominal open-loop trajectory is given by dash-dotted line and **✗**. The tightened set $\overline{\mathbb{X}}$ is depicted by the dashed lines.

Problem 3.6.

$$\underset{z(0|t),\, c(t)}{\text{minimize}} \; J_N^{\max}\Big(z(0|t), \boldsymbol{c}(t)\Big)$$

subject to

$$z(k+1|t) = f_\pi\Big(z(k|t), c(k|t), 0\Big), \qquad \forall k \in \mathbb{I}_{[0,N-1]}, \qquad (3.17\text{a})$$

$$x(t) \in \Big\{z(0|t)\Big\} \oplus \Omega, \qquad\qquad\qquad (3.17\text{b})$$

$$\Big(z(k|t), c(k|t)\Big) \in \overline{\mathbb{Z}} \qquad\qquad \forall k \in \mathbb{I}_{[0,N-1]}, \qquad (3.17\text{c})$$

$$z(N|t) \in \overline{\mathbb{X}}_{\mathrm{f}}, \qquad\qquad\qquad (3.17\text{d})$$

where

$$J_N^{\max}\Big(z(0|t), \boldsymbol{c}(t)\Big) = \sum_{k=0}^{N-1} \max_{\omega \in \Omega} \ell_\pi\Big(z(k|t)+\omega, c(k|t)\Big) + \overline{V}_{\mathrm{f}}\Big(z(N|t)\Big). \qquad (3.18)$$

The main idea in this section is highlighted in the objective (3.18). The maximization is "shifted" into the sum in the objective considering the worst-case error possible inside the RCI set Ω. In contrast to Problem 3.1, the disturbance is no longer an optimization variable. In fact it does not even occur in the optimization problem. Instead, the nominal dynamics are considered and the nominal initial state $z(0|t)$ is an additional optimization variable. Moreover, we take the input π into account within the objective, and thus, also contribute the input which is applied to the system in order to keep the system state within the RCI set. This is in contrast to the approaches considered in tube-based robust stabilizing MPC, where usually only the nominal input is considered within the stage cost function. The optimal nominal initial state is denoted by $z^*(0|t)$. Furthermore, we denote the value function of this problem by $V_N^{\max}(x(t))$, the minimizing input sequence by

$\boldsymbol{c}^*(t) = \{c^*(0|t), \dots, c^*(N-1|t)\}$, and the associated nominal state sequence satisfying (3.14) by $\boldsymbol{z}^*(t) = \{z^*(0|t), \dots, z^*(N|t)\}$.

For the problem setup in Problem 3.6, we can introduce the robust counterpart of the optimal steady-state (z_s^c, c_s^c) defined in (3.11) by

$$(z_s^{\max}, c_s^{\max}) = \underset{z=f_\pi(z,c,0),\, (x,c)\in\overline{\mathbb{Z}}}{\arg\min} \; \max_{\omega\in\Omega} \ell_\pi(z+\omega, c). \tag{3.19}$$

Again, existence follows by Assumptions 2.1–2.3 and 2.8 as well as continuity of the input parametrization. The difference between the two optimal steady-states (x_s^c, c_s^c) and $(z_s^{\max}, c_s^{\max})$ is in the objectives as well as in the constraint sets $\overline{\mathbb{Z}}$ and $\underline{\mathbb{Z}}$, respectively. While the optimal steady-state in the classical setup is determined with respect to the original stage cost function ℓ, the robust optimal steady-state in the min-max setup based on tubes is derived based on the worst-case cost. One can compare the constraint sets $\overline{\mathbb{Z}}$ and $\underline{\mathbb{Z}}$. By means of the following proposition, one can replace $\overline{\mathbb{Z}}$ by the (maybe conservative) estimate $\underline{\mathbb{Z}}$ in (3.11):

Proposition 3.7. *Let Ω be an RCI set for the error dynamics (3.15), and let $\overline{\mathbb{Z}}_s$ and $\underline{\mathbb{Z}}_s$ be the sets of all steady-states of the nominal system (3.14) in $\overline{\mathbb{Z}}$ and $\underline{\mathbb{Z}}$, respectively, that is, $\overline{\mathbb{Z}}_s := \{(x,c) \in \overline{\mathbb{Z}} : x = f_\pi(x,c,0)\}$ and $\underline{\mathbb{Z}}_s := \{(x,c) \in \underline{\mathbb{Z}} : x = f_\pi(x,c,0)\}$. Then it holds that $\underline{\mathbb{Z}}_s \subseteq \overline{\mathbb{Z}}_s$.*

The following assumption is imposed on the terminal ingredients (terminal cost, terminal region, and terminal controller) of Problem 3.6:

Assumption 3.2. *There exists a terminal region $\overline{\mathbb{X}}_f \subseteq \mathbb{X}$, containing the robust optimal steady-state $z_s^{\max}$ in its interior, an auxiliary controller $\bar{\kappa}_f : \overline{\mathbb{X}}_f \to \mathbb{R}^m$, and a terminal cost $\overline{V}_f : \overline{\mathbb{X}}_f \to \mathbb{R}$ such that for all $z \in \overline{\mathbb{X}}_f$ it holds that*

(i) $\left(z, \bar{\kappa}_f(z)\right) \in \overline{\mathbb{Z}}$,

(ii) $f_\pi\left(z, \bar{\kappa}_f(z), 0\right) \in \overline{\mathbb{X}}_f$,

(iii) $\overline{V}_f\left(f_\pi(z, \bar{\kappa}_f(z), 0)\right) - \overline{V}_f(z) \leq -\max_{\omega\in\Omega} \ell_\pi\left(z + \omega, \bar{\kappa}_f(z)\right) + \max_{\omega\in\Omega} \ell_\pi\left(z_s^{\max} + \omega, c_s^{\max}\right)$.

Moreover, the terminal cost $\overline{V}_f$ is continuous on $\overline{\mathbb{X}}_f$.

Similar as in Assumption 2.6, (i) guarantees feasibility of the terminal auxiliary controller, and (ii) guarantees invariance of the terminal region under the terminal controller. Considering the terminal cost function, we note that in condition (iii) the maximization on the right hand side is only taken with respect to the first argument, i.e., $\bar{\kappa}_f(z)$ is independent of ω. The terminal controller is the input to the nominal system after $N-1$ prediction steps, and therefore, it is independent of the disturbances acting on the system. Considering the relation of the terminal region and the robust optimal steady-state, we recall Remark 2.23, even though stability of the min-max approach based on tubes is only investigated in the next chapter (see Section 4.3).

The following algorithm provides the controller in the min-max setup:

Algorithm 3.8 (Min-max robust economic model predictive control based on tubes). *At each time instant $t \in \mathbb{I}_{\geq 0}$, measure the state $x(t)$ and solve Problem 3.6. Apply the control input $u(t) := \pi(x(t), c^*(0|t))$ to system (3.1).*

With this algorithm, we can state the main result of this section:

Theorem 3.9. *Let $x_0 \in \overline{\mathbb{X}}_N \oplus \Omega$. Suppose that Assumptions 2.1, 2.3, 2.7–2.9, and 3.2 are satisfied and let $(z_s^{\mathrm{max}}, c_s^{\mathrm{max}})$ be the optimal steady-state according to (3.19). Then, Problem 3.6 is feasible for all $t \in \mathbb{I}_{\geq 0}$. Furthermore, the closed-loop system resulting from applying Algorithm 3.8 satisfies the pointwise-in-time constraints (2.2) for all $t \in \mathbb{I}_{\geq 0}$ as well as the asymptotic average performance bound*

$$\limsup_{T \to \infty} \frac{1}{T} \sum_{t=0}^{T-1} \ell_\pi\Big(x(t), c^*(0|t)\Big) \leq \max_{\omega \in \Omega} \ell_\pi\Big(z_s^{\mathrm{max}} + \omega, c_s^{\mathrm{max}}\Big). \qquad (3.20)$$

An interpretation for the right hand side of the asymptotic average performance result stated in (3.20) is in order: There does not exist a steady-state of the real system due to unknown disturbances acting on the system at each time instance. However, the set $\{z_s^{\mathrm{max}}\} \oplus \Omega$ can be interpreted as a robust counterpart of the steady-state. When applying the manipulated input c_s^{max} at each time $t \in \mathbb{I}_{\geq 0}$, for any initial state x_0 such that $x_0 \in \{z_s^{\mathrm{max}}\} \oplus \Omega$ it follows that $x(t) \in \{z_s^{\mathrm{max}}\} \oplus \Omega$ for all $t \in \mathbb{I}_{\geq 0}$. Thus, the right hand side of (3.20) is the worst-case stage cost when operating the system close to the optimal steady-state with the nominal input c_s^{max}.

Remark 3.10. *As discussed in Remark 2.18, allowing the initial nominal state $z(0|t)$ as an additional optimization variable in general provides better performance results, however, the closed-loop sequence of nominal states, i.e., $\{z^*(0|0), z^*(0|1), \dots\}$ does not satisfy the nominal dynamics (3.14). Thus, the sequence of nominal initial states can be fixed to satisfy the nominal dynamics by replacing (3.17b) with*

$$z(0|t) = f\Big(z(0|t-1), v^*(0|t-1), 0\Big), \quad \forall t \in \mathbb{I}_{\geq 1}, \qquad (3.21)$$

and an arbitrary $z(0|0)$ such that $x_0 \in \{z(0|0)\} \oplus \Omega$. We will see later that this more restrictive and more conservative constraint can simplify the analysis significantly (Section 4.2.1). Interestingly, even though the constraints as well as the algorithm are modified, the bound on the asymptotic average performance is, again, (3.20). However in closed-loop, the performance achieved by fixing the nominal initial state by means of (3.21) is usually worse (see Section 6.3.2 and 6.3.5).

3.2.1 Finding an appropriate quadratic terminal cost

The result on the bound of the asymptotic average performance in Theorem 3.9 is based on Assumption 3.2(iii), that is, on finding an appropriate terminal cost $\overline{V}_f$. In this section, we provide a constructive approach to derive a quadratic terminal cost which can be applied within the min-max robust economic MPC approach based on tubes. Restricting the terminal cost to be a quadratic function can lead to a (slight) deterioration of the performance bound in some cases. We introduce the following assumption and (without loss of generality) set $(z_s^{\mathrm{max}}, c_s^{\mathrm{max}}) = (0, 0)$:

Assumption 3.3. *(i) The function f_π is twice continuously differentiable on $\mathbb{X} \times \mathbb{U}$. (ii) The linearized nominal system $z(t+1) = Az(t) + Bc(t)$, with $A := \frac{\partial}{\partial x} f_\pi(0, 0, 0)$ and $B := \frac{\partial}{\partial c} f_\pi(0, 0, 0)$, is stabilizable.*

In the following, we choose the terminal controller $\bar{\kappa}_f(z(t)) = Kz(t)$, with $K \in \mathbb{R}^{m \times n}$, such that the origin is asymptotically stable for the linear closed-loop system $z(t+1) = A_{cl}z(t)$, where $A_{cl} = A + BK$.[1] The basic idea of this approach follows Amrit et al. (2011), where a methodology is presented to construct an appropriate terminal cost for a nominal economic MPC setup. We want to find a quadratic function of the form

$$\bar{\ell}(z) = \frac{1}{2}z^{\top}\overline{Q}z + q^{\top}z + \bar{c}, \tag{3.22}$$

with $\overline{Q} = \overline{Q}^{\top}$ such that for all $z \in \overline{\mathbb{X}}_f$ it satisfies

$$\bar{\ell}(z) \geq \max_{\omega \in \Omega} \ell_{\pi}(z + \omega, Kz) - \max_{\omega \in \Omega} \ell_{\pi}(\omega, 0). \tag{3.23}$$

We present a quadratic terminal cost function which is designed for the linearized dynamics only, see Assumption 3.3(ii). In order to take the nonlinear part of the dynamics $(f_{\pi}(z, Kz, 0) - A_{cl}z)$ into account, robustness with respect to the nonlinearity must be introduced. Moreover, the matrix $\overline{Q}$ can in principle be indefinite, however, the presented procedure can only be applied to positive definite matrices. Thus, we set $Q := \overline{Q} + \alpha I_n \succ 0$, with $\alpha \geq 0$, and it holds that

$$\ell_q(z) = \frac{1}{2}z^{\top}Qz + q^{\top}z + \bar{c} \geq \bar{\ell}(z) + \frac{\alpha}{2}z^{\top}z, \qquad \forall z \in \overline{\mathbb{X}}_f. \tag{3.24}$$

In order to determine α, a method is presented in Amrit et al. (2011) (employing Assumption 3.3(i)). Using ℓ_q from (3.24), we introduce the terminal cost candidate function $\overline{V}_q$ by

$$\overline{V}_q(z) := \frac{1}{2}z^{\top}Pz + p^{\top}z, \tag{3.25}$$

where P is the solution of the Lyapunov equation

$$P = A_{cl}^{\top}PA_{cl} + Q \tag{3.26}$$

and

$$p^{\top} = q^{\top}(I_n - A_{cl})^{-1}. \tag{3.27}$$

It follows that

$$\overline{V}_q\Big(f_{\pi}(z, Kz, 0)\Big) - \overline{V}_q(z) \leq -\ell_q(z) + \bar{c} + \frac{\alpha}{2}z^{\top}z \leq -\bar{\ell}(z) + \bar{c}. \tag{3.28}$$

This holds due to the fact that α is chosen such that it overestimates the nonlinearities occurring in $\overline{V}_q(f_{\pi}(z, Kz, 0)) - \overline{V}_q(A_{cl}z)$.

Applying this terminal cost in Problem 3.6 with Algorithm 3.8, an asymptotic average performance bound similar to (3.20) follows, namely

$$\limsup_{T \to \infty} \frac{1}{T}\sum_{t=0}^{T-1} \ell_{\pi}\Big(x(t), c^*(0|t)\Big) \leq \max_{\omega \in \Omega} \ell_{\pi}\Big(\omega, 0\Big) + \bar{c}. \tag{3.29}$$

---

[1]For many systems with stabilizing π, e.g., stabilizable linear systems with $\pi(x, c) = Kx + c$, choosing the terminal controller to zero, i.e., $\bar{\kappa}_f(x) \equiv 0$, can be a feasible choice. This can simplify the search for a terminal region significantly.

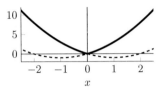

Figure 3.2: Right hand side of (3.23) for $\ell_\pi(x,c) = x^2 - 1$ with $\Omega = [-1,1]$, i.e., $\max_{\omega \in [-1,1]}(x+\omega)^2 - 1$.

By definition of $\bar{\ell}$ and by (3.23), $\bar{c} \geq 0$. We can see that the additional offset $\bar{c} \geq 0$ introduced in the quadratic function (3.22) appears in the asymptotic average performance bound as well. By this additional parameter, the provable performance bound is deteriorated.

It remains to discuss when $\bar{c} > 0$ is useful or even necessary for finding a quadratic terminal cost. The need for the constant $\bar{c}$ is directly related to the differentiability of the stage cost function. In Amrit et al. (2011), the stage cost is assumed to be (at least) twice continuously differentiable ($\mathcal{C}^2$). With this additional assumption, one can always find an appropriate $\overline{Q}$ and q such that (3.23) is satisfied with $\bar{c} = 0$. However, the right hand side of (3.23) need not be $\mathcal{C}^2$ in general. This can be seen when investigating the simple example $\ell_\pi(x,c) = x^2 - 1$ with $\Omega = [-1,1]$. Obviously, ℓ_π is twice continuously differentiable. However, when computing the right hand side of (3.23) (depicted in Figure 3.2), this is not $\mathcal{C}^2$ at the origin. For this example, $\bar{c} > 0$ is needed in order to be able to find an $\bar{\ell}$ satisfying condition (3.23).

On the one hand, from a performance point of view it seems intuitive at first sight that $\bar{c} \geq 0$ should be as small as possible. On the other hand, the three parameters $\overline{Q}$, q, and $\bar{c}$ cannot be chosen independently. Usually, choosing $\bar{c}$ smaller implies a larger $\overline{Q}$. This induces a larger impact of the terminal cost on the objective (3.18), since it puts a higher weight on the terminal cost. Thus, there is a trade-off between minimizing the performance bound (3.29) and the impact of the terminal cost. This trade-off could, for example, be handled with an optimization problem.

Example 3.11. *We consider a scalar linear system $x^+ = 0.5 + u + w$ with the constraints on the states $\mathbb{X} = [-5,5]$, on the inputs $\mathbb{U} = [-2,2]$, and the disturbance set $\mathbb{W} = [-0.1, 0.1]$. The stage cost is given by $\ell(x,u) = \frac{1}{2}x^2 + \frac{1}{2}u^2$ and the input parametrization by $\pi(x,c) = c$. The optimal steady-state is at the origin, and for the terminal controller, we use $\bar{\kappa}_f(x) \equiv 0$. As indicated by the sketch in Figure 3.2, the resulting stage cost is not $\mathcal{C}^2$ in x, thus $\bar{c} > 0$ is required. Some feasible choices are, for example, $\overline{Q} = 2$ with $\bar{c} = 0.015$ as well as $\overline{Q} = 10$ with $\bar{c} = 0.0022$. Here, $q = 0$ due to symmetry. These quadratic functions lead to a terminal cost with $P = 2.667$ (for $\overline{Q} = 2$) and $P = 13.333$ (for $\overline{Q} = 10$), respectively.*

Up to now, we have not discussed how to find a terminal region satisfying the conditions in Assumption 3.2. In Amrit et al. (2011), the terminal region is designed along with the terminal cost. This idea can be applied in our setup as well, since Assumption 3.2 is stated in terms of the nominal dynamics (3.14) only. For linear systems with a linear stabilizing input parametrization, the terminal region can be chosen to be the maximal output admissible set (see, e.g., Kolmanovsky and Gilbert (1998)).

3.2.2 Economic MPC in min-max setup based on tubes for linear system

The approach presented in the previous section is based on nonlinear dynamics and finding an appropriate RCI set in order to take all possible disturbances into account. However, as already described in Section 2.2, for linear systems less conservative tubes can be found. We consider linear time-invariant systems of the form

$$x(t+1) = Ax(t) + Bu(t) + w(t), \qquad x(0) = x_0, \tag{3.30}$$

satisfying Assumption 2.10 together with the linear input parametrization (3.8), i.e., $\pi(x, c) = Kx + c$, such that $A_{\mathrm{cl}} = A + BK$ is asymptotically stable (in the sense that all eigenvalues are strictly inside the unit disc). Following the approach in Chisci et al. (2001), the error (compare to (2.22)) between the linear system and its associated nominal system evolves within the time-varying sets

$$\Omega_{t+1} = A_{\mathrm{cl}}\Omega_t \oplus \mathbb{W}, \qquad \Omega_0 = \{x_0 - z_0\}. \tag{3.31}$$

These tubes are still over-approximations, yet, they are tighter than using the RCI set. As discussed in Section 2.2, one possible RCI set for this setup is the mRPI set Ω_∞, also representing the limit of (3.31) as $t \to \infty$. Using the constraint tightening introduced in (2.25), the resulting sets can be employed to determine a min-max robust economic receding horizon control law based on tubes similar to the one determined by Problem 2.19:

Problem 3.12.

$$\operatorname*{minimize}_{\mathbf{c}(t)} J_N^{\max}\Big(x(t), \mathbf{c}(t)\Big)$$

subject to

$$z(k+1|t) = A_{\mathrm{cl}}z(k|t) + Bc(k|t), \qquad \forall k \in \mathbb{I}_{[0,N-1]}, \tag{3.32a}$$
$$z(0|t) = x(t), \tag{3.32b}$$
$$(z(k|t), Kz(k|t) + c(k|t)) \in \overline{\mathbb{Z}}_k, \qquad \forall k \in \mathbb{I}_{[0,N-1]}, \tag{3.32c}$$
$$z(N|t) \in O_{\max} \ominus \Omega_N, \tag{3.32d}$$

where

$$J_N^{\max}\Big(x(t), \mathbf{c}(t)\Big) = \sum_{k=0}^{N-1} \max_{\omega \in \Omega_k} \ell_\pi\Big(z(k|t) + \omega, c(k|t)\Big) + \tilde{V}_{\mathrm{f}}\Big(z(N|t)\Big). \tag{3.33}$$

Considering the terminal constraint (3.32d), we recall that $O_{\max}$ denotes the maximal output admissible set with respect to $(z_s^{\max}, c_s^{\max})$ and $\mathbb{W}$, see Section 2.2.3.

The main advantage of this approach compared to Problem 3.6 lies in the objective (3.33) and the constraint tightening (3.32c). Instead of maximizing the stage cost over the whole mRPI set, we only take the corresponding tube into account which varies with respect to the prediction time k. Thus, fewer states are considered at each stage in the objective. For a geometric interpretation, consider again Figure 3.1: The sets Ω_k are less restrictive (there denoted by X_k) than the over-approximation Ω.

Due to the different structure of the optimal control problem, we also have to consider different conditions on the terminal ingredients:

Assumption 3.4. *Given the auxiliary controller* $\tilde{\kappa}_f(z) = c_s^{\max}$ *and a robust invariant terminal region* $\overline{\mathbb{X}}_f = O_{\max} \ominus \Omega_N$, *the terminal cost* $\tilde{V}_f : \overline{\mathbb{X}}_f \to \mathbb{R}$ *is continuous on* $\overline{\mathbb{X}}_f$ *and satisfies for all* $z \in \overline{\mathbb{X}}_f$ *and for all* $w \in \mathbb{W}$

$$\tilde{V}_f(A_{cl}z + Bc_s^{\max} + A_{cl}^N w) - \tilde{V}_f(z) \leq -\max_{\omega \in \Omega_{N-1}} \ell_\pi(z + A_{cl}^{N-1}w + \omega, c_s^{\max}) + \max_{\omega \in \Omega_\infty} \ell_\pi(z_s^{\max} + \omega, c_s^{\max}),$$
(3.34)

where $(z_s^{\max}, c_s^{\max})$ *is the optimal steady-state according to* (3.19).

The condition on the terminal cost function clarifies when computing the difference of the value function of Problem 3.12 at two consecutive time instances, similar as, for example, in the proof of Theorem 3.9. The additional disturbance $w \in \mathbb{W}$ required stems from (3.32b), that is, the constraint on the nominal initial state. Since the nominal initial state $z(0|t+1)$ is set to the measured real system state at time $t+1$, we have to consider $z(0|t+1) = A_{cl}x(t) + Bc^*(0|t) + w(t)$ such that the additional disturbance term emerges.

Remark 3.13. *In principle it is possible to choose a terminal controller different to the one in Assumption 3.4, for example, one could choose a linear feedback of the form* $\tilde{\kappa}_f(z) = K_f z$, *where* $K_f \in \mathbb{R}^{m \times n}$ *such that* $A + B(K + K_f)$ *is stable. In this case, one would need to determine the terminal region differently in order to provide robust invariance with this new controller.*

We will not go into detail with the analysis of this approach since most of it has already been presented for the previous method based on over-approximation by means of RCI sets (see Proof of Theorem 3.9). Even though this approach appears to be advantageous, it turns out that the provable average performance bound is, again, (3.20). This is due to the problem of stating performance bounds a priori including the unknown and unpredictable disturbance sequence. Thus, for the performance bound, the average performance is related to Ω_∞ (see discussion after Theorem 3.9).

Finding an appropriate quadratic terminal cost

We aim at providing an appropriate quadratic terminal cost satisfying Assumption 3.4. It turns out that for this setup it is more involved than the procedure presented in Section 3.2.1. Again, using a quadratic terminal cost may lead to a deterioration of the provable performance bound. The general idea is similar to the one presented in Section 3.2.1. One searches for a quadratic terminal cost satisfying (3.34) with a quadratic approximation of the respective right hand side and apply this new terminal cost when computing performance bound. Again, without loss of generality, we set $(z_s^{\max}, c_s^{\max}) = (0, 0)$.

We consider

$$\bar{\ell}(z) = \max_{\omega \in \Omega_N} \ell(z + \omega, K(z + \omega)) - \max_{\omega \in \Omega_N} \ell(\omega, K\omega),$$
(3.35)

where the second term is constant. We highlight that both terms take the maximum over Ω_N into account. This is different to the right hand side of (3.34), where the second term maximizes with respect to the mRPI set Ω_∞. This is in order to guarantee that $\bar{\ell}(0) = 0$.

Next, we over-approximate $\bar{\ell}$ by a quadratic function satisfying

$$\ell_q(z) = \frac{1}{2}z^\top Q z + q^\top z + \bar{c} \geq \bar{\ell}(z), \quad \forall z \in \overline{\mathbb{X}}_f,$$
(3.36)

with $Q = Q^\top \succeq 0$. The possible need for the additional constant $\bar{c} \geq 0$ has been discussed in Section 3.2.1. Using this approximation, a quadratic terminal cost candidate follows by

$$\tilde{V}_{\mathrm{q},\varepsilon}(z) = \frac{1}{2}z^\top P_\varepsilon z + p^\top z, \tag{3.37}$$

where P_ε is the solution of the Lyapunov equation

$$P_\varepsilon = A_{\mathrm{cl}}^\top P_\varepsilon A_{\mathrm{cl}} + (Q + \varepsilon I_n), \tag{3.38}$$

with $\varepsilon > 0$, and p is the same as in (3.27). Computing the difference of the candidate terminal cost at two consecutive steps employing the terminal controller, it follows for all $w \in \mathbb{W}$ that

$$
\begin{aligned}
&\tilde{V}_{\mathrm{q},\varepsilon}(A_{\mathrm{cl}}z + A_{\mathrm{cl}}^N w) - \tilde{V}_{\mathrm{q},\varepsilon}(z) \\
&= \frac{1}{2}(A_{\mathrm{cl}}z + A_{\mathrm{cl}}^N w)^\top P_\varepsilon (A_{\mathrm{cl}}z + A_{\mathrm{cl}}^N w) + p^\top (A_{\mathrm{cl}}z + A_{\mathrm{cl}}^N w) - \frac{1}{2}z^\top P_\varepsilon z - p^\top z \\
&\leq -\ell_{\mathrm{q}}(z) + \bar{c} - \frac{1}{2}z^\top (\varepsilon I_n) z + (A_{\mathrm{cl}}^N w)^\top P_\varepsilon A_{\mathrm{cl}} z + \frac{1}{2}(A_{\mathrm{cl}}^N w)^\top P_\varepsilon A_{\mathrm{cl}}^N w + p^\top A_{\mathrm{cl}}^N w \\
&\leq -\max_{\omega \in \Omega_N} \ell\big(z + \omega, K(z + \omega)\big) + \max_{\omega \in \Omega_N} \ell(\omega, K\omega) \\
&\quad + \bar{c} - \frac{1}{2}z^\top(\varepsilon I_n)z + (A_{\mathrm{cl}}^N w)^\top P_\varepsilon A_{\mathrm{cl}} z + \frac{1}{2}(A_{\mathrm{cl}}^N w)^\top P_\varepsilon A_{\mathrm{cl}}^N w + p^\top A_{\mathrm{cl}}^N w \\
&\leq -\max_{\omega \in \Omega_{N-1}} \ell\big(z + A_{\mathrm{cl}}^{N-1}w + \omega, K(z + A_{\mathrm{cl}}^{N-1}w + \omega)\big) + \max_{\omega \in \Omega_N} \ell(\omega, K\omega) \\
&\quad + \bar{c} - \frac{1}{2}z^\top(\varepsilon I_n)z + (A_{\mathrm{cl}}^N w)^\top P_\varepsilon A_{\mathrm{cl}} z + \frac{1}{2}(A_{\mathrm{cl}}^N w)^\top P_\varepsilon A_{\mathrm{cl}}^N w + p^\top A_{\mathrm{cl}}^N w.
\end{aligned}
\tag{3.39}
$$

A brief discussion considering the derivation is in order: Due to (3.32b) and z representing $z(N|t)$ of the open-loop trajectory, we have to consider the disturbance acting on the system at time t. Hence, this disturbance appears as $A_{\mathrm{cl}}^N w$ and our sole knowledge about this disturbance is $w \in \mathbb{W}$. The first inequality follows by the definition of P_ε and p, the second inequality by (3.36). The third inequality follows since $\max_{\omega \in \Omega_N} \ell(z + \omega, K(z + \omega)) \geq \max_{\omega \in \Omega_{N-1}} \ell(z + A_{\mathrm{cl}}^{N-1}w + \omega, K(z + A_{\mathrm{cl}}^{N-1}w + \omega))$ for all $w \in \mathbb{W}$. In order to provide meaningful—i.e. small—performance bounds, as many terms as possible should be canceled. Therefore, we consider the mixed term $(A_{\mathrm{cl}}^N w)^\top P_\varepsilon A_{\mathrm{cl}} z$ satisfying

$$
\begin{aligned}
(A_{\mathrm{cl}}^N w)^\top P_\varepsilon A_{\mathrm{cl}} z &\leq \left| (A_{\mathrm{cl}}^N w)^\top P_\varepsilon A_{\mathrm{cl}} z \right| \leq \sqrt{(A_{\mathrm{cl}}^N w)^\top P_\varepsilon A_{\mathrm{cl}}^N w}\sqrt{(A_{\mathrm{cl}}z)^\top P_\varepsilon A_{\mathrm{cl}} z} \\
&\leq \frac{\bar{\varepsilon}}{2}(A_{\mathrm{cl}}^N w)^\top P_\varepsilon A_{\mathrm{cl}}^N w + \frac{1}{2\bar{\varepsilon}}(A_{\mathrm{cl}}z)^\top P_\varepsilon A_{\mathrm{cl}} z,
\end{aligned}
\tag{3.40}
$$

for every $\bar{\varepsilon} > 0$, where the second inequality follows by Cauchy-Schwarz and the third by using Young's inequality. Choosing $\bar{\varepsilon}$ such that $\frac{1}{\bar{\varepsilon}}A_{\mathrm{cl}}^\top P_\varepsilon A_{\mathrm{cl}} - \varepsilon I_n \preceq 0$, it follows that

$$
\begin{aligned}
&\tilde{V}_{\mathrm{q},\varepsilon}(A_{\mathrm{cl}}z + A_{\mathrm{cl}}^N w) - \tilde{V}_{\mathrm{q},\varepsilon}(z) \\
&\leq -\max_{\omega \in \Omega_{N-1}} \ell\big(z + A_{\mathrm{cl}}^{N-1}w + \omega, K(z + A_{\mathrm{cl}}^{N-1}w + \omega)\big) + \max_{\omega \in \Omega_N} \ell(\omega, K\omega) \\
&\quad + \bar{c} + \frac{1 + \bar{\varepsilon}}{2}(A_{\mathrm{cl}}^N w)^\top P_\varepsilon A_{\mathrm{cl}}^N w + p^\top A_{\mathrm{cl}}^N w.
\end{aligned}
\tag{3.41}
$$

Applying the same steps as in the proof of Theorem 3.9, we can derive for the asymptotic average performance

$$
\begin{aligned}
\limsup_{T \to \infty} &\frac{1}{T} \sum_{t=0}^{T-1} \ell\big(x(t), Kx(t) + c^*(0|t)\big) \\
&\leq \max_{\omega \in \Omega_N} \ell\big(\omega, K\omega\big) + \bar{c} + \max_{w \in \mathbb{W}} \left\{ \frac{1+\bar{\varepsilon}}{2} (A_{\mathrm{cl}}^N w)^\top P_\varepsilon A_{\mathrm{cl}}^N w + p^\top A_{\mathrm{cl}}^N w \right\}.
\end{aligned}
\tag{3.42}
$$

The performance bound stated in (3.42) is usually worse than the bound in (3.29). Taking a close look at the bound in (3.42), one can see, however, that this is an approximation for the bound in (3.29) in the following sense: The bound in (3.29) accounts for the worst case over all possible disturbances in the RCI set centered at the optimal steady-state. Since linear systems are considered in this section, the RCI set can be assumed to be the mRPI set. Each state within the mRPI set can be described by a (possibly infinite) sum of disturbances, i.e., for each $\omega \in \Omega_\infty$ there exists a sequence $\boldsymbol{w} = \{w(0), w(1), \dots\}$ with $w(k) \in \mathbb{W}, \forall k \in \mathbb{I}_{\geq 0}$, such that $\omega = \sum_k A_{\mathrm{cl}}^k w(k)$. Focusing on (3.42), the bound is split into two parts. The first term maximizes over Ω_N, and hence, contributes for the first N disturbances in the (infinite) sequence. The third term provides a quadratic approximation of the disturbances for all times $k \in \mathbb{I}_{>N}$. This follows by construction of P_ε. However, this bound is conservative in two ways: First, the quadratic approximation is provided for (3.35), that is, with respect to $\max_{\omega \in \Omega_N} \ell(z + \omega, K(z + \omega))$. Second, the two "robustifying" parameters ε and $\bar{\varepsilon}$ are required incorporating the mixed term of the disturbance and the terminal state. Thus, one can see that using a quadratic terminal cost deteriorates the performance bound to a quadratic approximation of the performance derived with the general Assumption 3.4(iii). We will see similar results later when additional stochastic information is provided in Section 5.2.

Remark 3.14. *When regarding the performance bound in (3.42), it would be desirable to choose ε and $\bar{\varepsilon}$ such that the bound is as small as possible. Therefore, we note that due to linearity the solution of the Lyapunov equation (3.38) can be split into two parts $P_\varepsilon = P_Q + \varepsilon P_I$, where $P_Q = A_{\mathrm{cl}}^\top P_Q A_{\mathrm{cl}} + Q$ and $P_I = A_{\mathrm{cl}}^\top P_I A_{\mathrm{cl}} + I_n$. With this, the best ε and $\bar{\varepsilon}$ can be found by the argument of the minimization*

$$
\begin{aligned}
&\underset{\varepsilon, \bar{\varepsilon}}{minimize} \left\{ \underset{w \in \mathbb{W}}{maximize} \; \frac{1+\bar{\varepsilon}}{2} (A_{\mathrm{cl}}^N w)^\top P_Q (A_{\mathrm{cl}}^N w) + \varepsilon \frac{1+\bar{\varepsilon}}{2} (A_{\mathrm{cl}}^N w)^\top P_I (A_{\mathrm{cl}}^N w) \right\} \\
&subject \; to \quad \varepsilon > 0, \; \bar{\varepsilon} > 0, \\
&\qquad \frac{1}{\bar{\varepsilon}} A_{\mathrm{cl}}^\top P_Q A_{\mathrm{cl}} + \frac{\varepsilon}{\bar{\varepsilon}} A_{\mathrm{cl}}^\top P_I A_{\mathrm{cl}} - \varepsilon I_n \prec 0.
\end{aligned}
\tag{3.43}
$$

This problem is nonlinear, but approximate solutions can be found.

Remark 3.15. *The underlying basic idea of considering estimates of the system state based on sets that vary with respect to the prediction time could as well be adapted to nonlinear systems, if an appropriate formulation for the growing sets can be found. One possible approach is presented in Limón et al. (2002a), where the evolution of the sets is determined with respect to the Lipschitz constant of the dynamics.*

3.3 Discussion comparing the approaches

In this section, we want to discuss the advantages and disadvantages of the min-max robust economic MPC approaches presented in the previous sections. Therefore, we consider the structural conservatism within the optimization problem, the computational complexity, as well as the conservatism of the performance bounds.

Structural conservatism

The classical min-max robust economic MPC approach presented in Section 3.1 is the least conservative of the presented approaches. This is due to considering only combinations of feasible state, input, *and* disturbance sequences within the optimization. Moreover, the tightening of the constraints is only as tight as necessary reducing the feasible set only as little as necessary. With respect to simulations, this approach results in general in the best closed-loop average performance.

Taking the min-max robust economic MPC approaches based on tubes presented in Section 3.2 into account, the tubes introduce additional conservatism by representing the influence of the disturbance with invariant sets. When considering the resulting worst-case sequence $\boldsymbol{\xi}(t) = \{\xi(0|t), \xi(1|t), \dots, \xi(N-1|t)\}$ with

$$\xi(k|t) := \underset{\zeta \in \{z(k|t)\} \oplus \Omega_\infty}{\arg\max} \ \ell_\pi\Big(\zeta, c(k|t)\Big), \tag{3.44}$$

this sequence might not be a possible trajectory of the real system (3.1), i.e., there need not exist a sequence of inputs $\boldsymbol{c}(t) = \{c(0|t), \dots, c(N-1|t)\}$ and disturbances $\boldsymbol{w}(t) = \{w(0|t), \dots, w(N-1|t)\}$ such that $\boldsymbol{\xi}(t)$ is a solution of (3.1). When allowing the nominal initial state $z(0|t)$ as an additional optimization variable, see Problem 3.6, the additional degree of freedom can compensate for this conservatism, but still the closed-loop performance is in general worse than for the classical approach. (It is even worse when the nominal initial state is fixed to follow a trajectory, see Remark 3.10.)

Allowing for tubes varying with the prediction-time for linear systems presented in Section 3.2.2 can reduce the conservatism (see also Remark 3.15). This approach considers only errors that are possible along the open-loop predictions, and thus, reduces the tightening. However, there is still no guarantee that the resulting worst-case sequence is a possible trajectory of system (3.30). In simulations, this approach in general produces closed-loop average performances in between those of the two approaches above.

Computational complexity

The major disadvantage of the classical approach is the enormous computational complexity. As already discussed in Section 2.2.1, allowing to optimize over general feedback policies is intractable. However, even the approach presented in Section 3.1, where the input is parametrized by some feedback law π, is computationally expensive. Especially the maximization over all possible disturbances is cumbersome and can in general not be performed efficiently. Moreover, computing the state evolution along the open-loop predictions in order to provide a tightening of the constraints is complex for nonlinear systems (see Limón et al. (2009) for a detailed discussion), but can be performed offline.

The min-max approaches based on tubes require to find robust invariant sets for the error dynamics (3.15). In general, this is a challenging task for nonlinear systems. However, there are several methods and it suffices to find a robust invariant outer approximation (see discussion in Section 2.2.2). Moreover, this computation is performed offline. Of course, using large over-approximations increases the structural conservatism. Considering the computational complexity in the optimization, this problem is simple compared to the classical approach since it requires the maximization to be solved separately for each prediction step. For some examples, e.g., for convex RCI sets and a convex stage cost function ℓ, the maximization might be solved analytically.

When taking linear systems into account, computing the tubes and the tightening is possible for all of the presented approaches. Again, the min-max approaches based on tubes are in general computationally more appealing.

Conservatism of the guaranteed performance bound

Comparing the bounds on the asymptotic average performance in (3.13) and (3.20) is interesting for the min-max robust economic MPC approaches, since they provide bounds for the average closed-loop performance of the *real* system. By design, the min-max approaches take the worst-case disturbance into account. Thus, the bounds provided are in general larger than the closed-loop average performance since the disturbance acting on the system need not be the worst case. However, the bounds might be additionally conservative by design choices.

For the classical approach, the size of the bound in (3.13) depends to a large extend on the bounds found for the terminal cost function. This bound in turn depends on the $\mathcal{K}_\infty$ function ρ from Assumption 3.1(iii). As discussed in Section 3.1, ideas from the input-to-state stability literature can be used to find a terminal cost and ρ. However, as it is common when working with comparison functions, the provided ρ can be quite conservative, which deteriorates the performance bound.

The quality of the bound in the min-max robust economic MPC based on tubes is mainly depending on the size of the RCI set. Finding a small—ideally minimal—RCI is a challenging task and directly affects not only the performance bound, but also the closed-loop performance in this setup since it appears within the objective (3.18). Additional conservatism might be introduced when using quadratic terminal cost functions, see Section 3.2.1. We recall that this conservatism can be traded off with the influence of the terminal cost.

For the linear min-max case based on varying tubes, the performance bound derived when using a quadratic terminal cost function is in general more conservative than the one based on mRPI sets (see discussion in Section 3.2.2). One major source of conservatism is the over-estimation of the mixed term in (3.40) by two quadratic terms. One can reduce this inherent conservatism by using the involved design parameters ε and $\bar{\varepsilon}$, but only to a certain extent. Moreover, long prediction horizons and small absolute values of the eigenvalues of A_{cl} help to reduce the conservatism as well, as will become apparent in the numerical example in Section 6.3.

A possible way to tighten the performance bounds for the min-max setups based on tubes could be to consider robust steady-states (see, e.g., Brunner et al., 2016; Raković and Barić, 2010; Raković et al., 2007). Up to know, the bounds were determined with respect to the RCI sets centered at the optimal steady-state. It could be beneficial—from a performance point of view—to find different robust invariant sets chosen with respect to the desired objective and use those as terminal regions.

3.4 Numerical examples

In this section, we apply the previously presented approaches for min-max robust economic MPC to two numerical examples. The first example is linear and scalar in order to provide some intuition about the performance bounds, the second example is a nonlinear example.

3.4.1 Linear example

As a first example, we investigate the performance of the approximating min-max approach based on tubes in comparison to the classical min-max approach. Therefore, we reconsider Example 3.11 given by the scalar linear system

$$x(t+1) = 0.5x(t) + u(t) + w(t) \tag{3.45}$$

with the constraints on the states $\mathbb{X} = [-5, 5]$, the inputs $\mathbb{U} = [-2, 2]$, and the bound on the disturbance $\mathbb{W} = [-0.1, 0.1]$. The stage cost function is given by

$$\ell(x, u) = \frac{1}{2}x^2 + \frac{1}{2}u^2. \tag{3.46}$$

The input parametrization $\pi(x, c) = c$ is used, which leads to $\Omega = [-0.2, 0.2]$; the prediction horizon is set to $N = 10$. The optimal steady-state (x_s^c, c_s^c) as well as the robust optimal steady-state $(z_s^{\max}, c_s^{\max})$ are at the origin.

Considering the classical min-max economic MPC approach, we employ the quadratic terminal cost $V_f(x) = \frac{1}{2}Px^2$ with $P = 2.1616$. The terminal cost is derived using the approach in Sontag and Wang (1995, Remark 2.4) and leads to a bound of $\rho(|w|) = 2.8637|w|^2$. Thus, for the performance bound follows $\rho(w_{\max}) = 0.0286$. For the min-max approach based on tubes, the performance bound results in $\max_{\omega \in \Omega} \ell(\omega, \pi(\omega, 0)) = 0.02$.

As mentioned in Example 3.11, for this example the worst-case stage cost is not $\mathcal{C}^2$ and a constant offset $\bar{c}$ is required in order to determine a terminal cost function for the tube-based approach. Following Example 3.11, we choose $P = 13.333$ in the following simulation which leads to a performance bound of 0.0222.

We receive the average closed-loop performance (averaged over 20 simulations starting at the origin with 200 closed-loop iterations each, and w uniformly distributed over $\mathbb{W}$) and the worst-case closed-loop performance (using the same number of iterations but applying the worst-case disturbance at each iteration):

	Algorithm 3.2	Algorithm 3.8
Performance bound	0.0286	0.0222
Average closed-loop performance	0.0019	0.0022
Worst-case closed-loop performance	0.0090	0.0200

We note that the performance bound is not tight for the average closed-loop performance. This is due to the fact that the performance bound takes the worst-case disturbance into account while the average stated in the table above is obtained from simulations where the disturbances are distributed uniformly over $\mathbb{W}$. When considering the worst-case disturbance in closed loop, one can see that the bound for the min-max approach based on tubes is tight in the sense that the average performance converges to 0.02, whereas the average performance of the classical approach is still better than the provided bound.

3.4.2 Nonlinear example

We now apply the presented approaches to a nonlinear example, namely to the bilinear system

$$x(t+1) = \begin{bmatrix} 0.55 & 0.12 \\ 0 & 0.67 \end{bmatrix} x(t) + \left(\begin{bmatrix} -0.6 & 1 \\ 1 & -0.8 \end{bmatrix} x(t) + \begin{bmatrix} 0.01 \\ 0.15 \end{bmatrix} \right) u(t) + w(t). \tag{3.47}$$

This model is slightly modified compared to the one originally introduced in Limón et al. (2002a), where the disturbance is multiplicative in the input. We also adapt the state and input constraints, such that $\mathbb{X} = \{x \in \mathbb{R}^n : \|x\|_\infty \leq 1\}$ and $\mathbb{U} = \{u \in \mathbb{R} : |u| \leq 0.15\}$. The disturbance set is modified as well and given by $\mathbb{W} = \{w \in \mathbb{R}^2 : \|w\|_\infty \leq 0.005\}$.

With regard to the min-max approach based on tubes, we need to determine an RCI set. For this system with the given constraints, one can make use of the framework presented in Bayer et al. (2013). The approach is based on finding an invariant level set of an appropriate incremental input-to-state stability Lyapunov function. By using the Lyapunov candidate function $V(e) = \|e\|_\infty$ for the error dynamics (2.15), an RCI set for these dynamics can be determined by

$$\Omega = \{e \in \mathbb{R}^2 : \|e\|_\infty \leq 5/60\}. \tag{3.48}$$

With this approach, the input for the real system is the same as for the nominal system, that is, $\pi(x,c) = c$. The cost function is given by

$$\ell(x,u) = 0.001\, x_2 + \begin{cases} -100\, x_1^3 & \text{for } x_1 < 0 \\ 0 & \text{for } x_1 \geq 0. \end{cases} \tag{3.49}$$

Using this cost function, we want to minimize the second state x_2 while keeping the first state x_1 positive by means of a soft constraint. Computing the optimal steady-state for this example, it follows $x_s^c = (0.0051, -0.1108)^\top$, the robust optimal steady-state is located at $z_s^{max} = (0.0333, 0.0611)^\top$. The prediction horizon is set to $N = 5$.

For the classical min-max approach, we use—as in the previous example—the concept from Sontag and Wang (1995, Remark 2.4) and a quadratic approximation of the stage cost in order to determine a quadratic terminal cost with $P = \begin{bmatrix} 451.468 & -235.703 \\ -235.703 & 57.405 \end{bmatrix}$ and $p = [-0.002, 0.005]^\top$ along with a $\mathcal{K}_\infty$-function ρ. Here, we employ the maximal output admissible set (under application of c_s^c) as the terminal region. The disturbance gain is determined to $\rho(|w|) = 379.4546|w|^2 + 0.0028|w|$ and results in a performance bound of $\rho(w_{max}) = 0.0190$. The performance bound for the min-max approach based on tubes is $\max_{\omega \in \Omega} \ell(x_s^{max} + \omega, c_s^{max}) = 0.0126$. The terminal region is derived as the maximal output admissible set under application of the terminal controller $\bar{\kappa}_f(x) = c_s^{max}$; the terminal cost is provided by the procedure in Section 3.2.1.

For the average closed-loop performance (averaged over 20 simulations starting at the origin with 200 closed-loop iterations each, and w uniformly distributed over $\mathbb{W}$) we receive:

Algorithm 3.2	Algorithm 3.8
$-10.7 \cdot 10^{-5}$	$6 \cdot 10^{-5}$

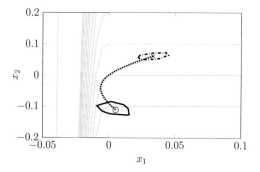

Figure 3.3: Contour plot of the stage cost ℓ in (3.49). All feasible steady-states are given by the dotted line. The convex hull of the closed-loop states after a settling time of 10 time steps are depicted by the sets (Algorithm 3.2 – solid; Algorithm 3.8 – dash-dotted). The optimal steady-states x_s^c and $z_s^{\max}$ are represented by the circle and cross, respectively.

Investigating the convex hull of the closed-loop states in Figure 3.3, it becomes apparent why the asymptotic average performance differs significantly. By design, the min-max approach based on tubes considers the worst case on the whole RCI set, which is large compared to the distance of all possible steady-states to the soft constraint $x_1 \geq 0$. Thus, the robust optimal steady-state $z_s^{\max}$ is chosen as far away from $x_1 < 0$ as possible and Algorithm 3.8 keeps the states of the closed-loop system in a neighborhood thereof. In contrast, the classical min-max approach takes the optimal steady-state x_s^c into account which is the best feasible nominal steady-state with respect to (3.49), i.e., without considering any influence of the disturbance within the stage cost. Algorithm 3.2 can keep the closed-loop system state in a neighborhood of x_s^c, due to the exact behavior of the disturbance along the prediction (see (3.9d)–(3.9f)), which results in a better asymptotic average performance.

3.5 Summary

In this chapter, we discussed first conceptual ideas to incorporate information provided about external disturbances within an economic MPC framework. Regarding the worst case over all disturbance sequences, the influence of the disturbance was taken into account within the optimization. The three presented approaches differ in the way the worst case is considered in the underlying optimal control problem. While the first setup allows for the worst case disturbance sequence within the objective, the latter setups approximate the influence by using ideas from tube-based robust MPC. By this, the worst case of the disturbance is not evaluated for open-loop disturbance sequences, but considered separately at each prediction step. These setups are dependent on the underlying (invariant) error sets. Constructive approaches were presented for finding appropriate terminal cost functions. In the classical approach, those are based on ideas from nonlinear control; for the approaches based on tubes, the terminal cost could be chosen quadratically, however, deteriorating the performance bound. For the presented setups, we discussed the structural conservatism induced into the optimization problems, the computational complexity as well as possible conservatism of the guaranteed performance bounds.

The main problem considering the performance bound is the usually unrealistic principle of min-max robust MPC, that is, to consider the worst-case disturbance within the prediction. This might be realistic for some setups (problems from game theory where the "opponent" chooses the disturbance), but for many practical applications, this is not the case. We present a similar approach being more applicable in case that the disturbance takes values on the whole disturbance set in the next chapter.

Chapter 4

Tube-based robust economic MPC

In this chapter, we present a second approach to consider disturbances in an economic MPC framework. As mentioned in the previous section on min-max robust economic MPC, always taking the worst case into account provides a robust statement. However, the results—especially the guaranteed performance bounds—are usually rather conservative in the min-max setup. In order to overcome this drawback, we present an economic MPC approach that considers the disturbance by averaging the stage cost over all possible errors. Since the evolution of the future errors can be approximated by invariant sets, we refer to this averaging approach as *tube-based robust economic MPC*.

For this new setup, we do not only investigate performance bounds (Section 4.1), but also optimal operating regimes (Section 4.2), as well as (asymptotic) stability (Section 4.3), and show connections to the min-max robust economic MPC setup in Chapter 3. Moreover, we take average constraints into account, that is, constraints that need not be satisfied at every time instance but only on average (Section 4.4).

The results presented in this chapter are based on (Bayer et al., 2014a,b, 2017) as well as on (Bayer and Allgöwer, 2014; Bayer et al., 2015b).

4.1 Problem setup and performance result

As in the previous chapters, we want to find a feasible input sequence to the nonlinear time-invariant disturbance-affected system

$$x(t+1) = f\Big(x(t), u(t), w(t)\Big), \qquad x(0) = x_0, \tag{4.1}$$

such that the asymptotic average cost

$$\limsup_{T \to \infty} \frac{1}{T} \sum_{t=0}^{T-1} \ell\Big(x(t), u(t)\Big) \tag{4.2}$$

is minimized while guaranteeing satisfaction of the pointwise-in-time constraints (2.2). The disturbance $w(t)$ satisfies Assumption 2.7. Again, the continuous stage cost function may be arbitrary, that is, it does not need to satisfy any assumption of the form of Assumption 2.5. Following the idea of tube-base robust stabilizing MPC presented in Section 2.2, we introduce an input parametrization equivalent to (2.14), that is,

$$u(t) = \varphi\Big(v(t), x(t), z(t)\Big),$$

where $z(t)$ is the state of and $v(t)$ is the input to the nominal system (2.12), respectively. Following Definition 2.13, we denote the robust control invariant error set by Ω, and with respect to the pointwise-in-time constraints, the constraints are tightened as in (2.16), providing the set $\overline{\mathbb{Z}} = \{(z,v) \in \mathbb{R}^n \times \mathbb{R}^m : (x, \varphi(v,x,z)) \in \mathbb{Z},\ \text{for all } x \in \{z\} \oplus \Omega\}$ satisfying Assumption 2.9.

The key idea in this chapter is to consider the disturbance by means of averaging the cost function over the RCI set. Thus, we can associate the disturbed system to the underlying nominal system while still accounting for (an over-approximation of) the influence of the disturbance. We introduce the integrated stage cost function ℓ^{int} defined as

$$\ell^{\text{int}}(z,v) := \int_{\omega \in \Omega} \ell\Big(z + \omega, \varphi(v, z + \omega, z)\Big)\mathrm{d}\omega. \tag{4.3}$$

By Assumption 2.8 this integral exists and is in general non-zero. Two main features differ from a standard tube-based (stabilizing) MPC approach as presented in Section 2.2.2. First, the integrated stage cost function considers all possible states within the RCI set around the nominal state and, with this, introduces robustness against disturbance in the economic objective. Second, as for the min-max approach based on tubes presented in Section 3.2, the "real" input φ, which is actually applied to the real system, is taken into consideration within the cost function. This means that instead of only considering the nominal input v, we also take the cost of the input into account which is needed to keep the real system state within the RCI set. This is in contrast to the aforementioned tube-based robust stabilizing MPC approaches, where usually only the nominal input is considered within the stage cost function.

We employ the integrated stage cost function within a finite-horizon optimization problem in order to determine a receding horizon control law for each $t \in \mathbb{I}_{\geq 0}$ given $x(t)$:

Problem 4.1.
$$\underset{z(0|t),\, \boldsymbol{v}(t)}{\text{minimize}}\ J_N^{\text{int}}\Big(z(0|t), \boldsymbol{v}(t)\Big)$$

subject to

$$z(k+1|t) = f\Big(z(k|t), v(k|t), 0\Big), \qquad\qquad \forall k \in \mathbb{I}_{[0,N-1]}, \tag{4.4a}$$

$$x(t) \in \{z(0|t)\} \oplus \Omega, \tag{4.4b}$$

$$\Big(z(k|t), v(k|t)\Big) \in \overline{\mathbb{Z}}, \qquad\qquad \forall k \in \mathbb{I}_{[0,N-1]}, \tag{4.4c}$$

$$z(N|t) \in \overline{\mathbb{X}}_{\text{f}}, \tag{4.4d}$$

where

$$J_N^{\text{int}}\Big(z(0|t), \boldsymbol{v}(t)\Big) = \sum_{k=0}^{N-1} \ell^{\text{int}}\Big(z(k|t), v(k|t)\Big) + V_{\text{f}}^{\text{int}}\Big(z(N|t)\Big). \tag{4.5}$$

With regard to the notation, we denote the value function by $V_N^{\text{int}}(x(t))$. The optimal nominal open-loop input and the associated nominal state sequences at time $t \in \mathbb{I}_{\geq 0}$ are denoted by $\boldsymbol{v}^*(t) = \{v^*(0|t),\dots,v^*(N-1|t)\}$ and $\boldsymbol{z}^*(t) = \{z^*(0|t),\dots,z^*(N|t)\}$, respectively.

Taking the integrated stage cost into account, we introduce the *robust optimal steady-state* by

$$(z_s^{\text{int}}, v_s^{\text{int}}) = \underset{z=f(z,v,0),\,(z,v)\in\overline{\mathbb{Z}}}{\arg\min}\ \ell^{\text{int}}(z,v). \tag{4.6}$$

Due to Assumptions 2.1, 2.3, 2.8, and 2.9, there exists a robust optimal steady-state, however, it need not be unique. If this is the case, one of the optimal steady-states can be chosen.

Considering the terminal region $\overline{\mathbb{X}}_f$ and the terminal cost $V_f^{\text{int}} : \overline{\mathbb{X}}_f \to \mathbb{R}$, we introduce the following assumption:

Assumption 4.1. *The terminal region $\overline{\mathbb{X}}_f \subseteq \overline{\mathbb{X}}$ is closed and contains z_s^{int} in its interior. There exists a local auxiliary controller $\bar{\kappa}_f : \overline{\mathbb{X}}_f \to \mathbb{U}$ such that for all $z \in \overline{\mathbb{X}}_f$ it holds that*

(i) $\left(z, \bar{\kappa}_f(z) \right) \in \overline{\mathbb{Z}},$

(ii) $f\left(z, \bar{\kappa}_f(z), 0 \right) \in \overline{\mathbb{X}}_f,$ *and*

(iii) $V_f^{\text{int}}\left(f(z, \bar{\kappa}_f(z), 0) \right) - V_f^{\text{int}}(z) \leq -\ell^{\text{int}}\left(z, \bar{\kappa}_f(z) \right) + \ell^{\text{int}}(z_s^{\text{int}}, v_s^{\text{int}}).$

Moreover, the terminal cost V_f^{int} is continuous on the terminal region $\overline{\mathbb{X}}_f$.

Similar to the min-max robust economic MPC scheme based on tubes in the previous section, the terminal cost V_f^{int} must guarantee a certain condition with respect to the integrated stage cost on the terminal region. If asymptotic stability shall not be investigated, Assumption 4.1 can be relaxed to satisfy $z_s^{\text{int}} \in \overline{\mathbb{X}}_f$ only (see Remark 2.23).

Under the additional assumption that ℓ, f, and φ (and hence also ℓ^{int}) are $\mathcal{C}^2$, and that the linearization at $(z_s^{\text{int}}, v_s^{\text{int}})$ is stabilizable (compare Assumption 3.3), the constructive method presented in Amrit et al. (2011, Section 4) can be used to determine a terminal cost V_f^{int} and a terminal region $\overline{\mathbb{X}}_f$ satisfying Assumption 4.1. In case that ℓ is not $\mathcal{C}^2$, one can follow the approach in Section 3.2.1.

Now, we can define the controller by means of the following algorithm:

Algorithm 4.2 (Tube-based robust economic model predictive control). *At each time instant $t \in \mathbb{I}_{\geq 0}$, measure the state $x(t)$ and solve Problem 4.1. Apply the control input $u(t) := \varphi\left(v^*(0|t), x(t), z^*(0|t) \right)$ to system (4.1).*

As for Problem 2.15 in the stabilizing setup, the nominal initial state $z(0|t)$ is an optimization variable. Thus, the closed-loop sequence of nominal states $\boldsymbol{z}^* = \{z^*(0|0), z^*(0|1), \dots\}$ equipped with the sequence of nominal inputs $\boldsymbol{v}^* = \{v^*(0|0), v^*(0|1), \dots\}$ need not be a trajectory for the nominal system (2.12). We refer to the Remarks 2.18 and 3.10 for more details. In the subsequent sections, it will be of interest to take two alternatives for determining the nominal initial state into account.

Using the presented algorithm, we state the main result of this section:

Theorem 4.3. *Let $x_0 \in \overline{\mathbb{X}}_N \oplus \Omega$. Suppose that Assumptions 2.1, 2.3, 2.7–2.9, and 4.1 are satisfied and let $(z_s^{\text{int}}, v_s^{\text{int}})$ be the robust optimal steady-state according to (4.6). Then, Problem 4.1 is feasible for all $t \in \mathbb{I}_{\geq 0}$. Furthermore, the closed-loop system resulting from Algorithm 4.2 satisfies the pointwise-in-time constraints (2.2) for all $t \in \mathbb{I}_{\geq 0}$. The closed-loop nominal sequence $\boldsymbol{z}^*$ with its associated input sequence $\boldsymbol{v}^*$ has a robust asymptotic average performance bound which is no worse than that of the robust optimal steady-state, i.e.,*

$$\limsup_{T \to \infty} \frac{1}{T} \sum_{t=0}^{T-1} \ell^{\text{int}}\left(z^*(0|t), v^*(0|t) \right) \leq \ell^{\text{int}}(z_s^{\text{int}}, v_s^{\text{int}}). \tag{4.7}$$

The proof for this theorem follows directly along the lines of the proof of Theorem 3.9 and is based on the tightening considered in $\overline{\mathbb{Z}}$. The average performance bound is guaranteed by the appropriate choice of the terminal cost in Assumption 4.1(iii).

An interpretation for the performance bound provided in (4.7) is in order: The statement (4.7) means that the average integral cost along the closed-loop nominal sequence z^* with the associated inputs v^* is at least as good as the integral cost at the robust optimal steady-state. As the states of the real closed-loop system resulting from applying Algorithm 4.2 lie in the RCI sets centered at the nominal sequence z^*, the performance result in (4.7) can be interpreted as an average performance result for the real closed-loop system averaged over all possible disturbances by integrating over the RCI set.

Remark 4.4. *Using the proposed integrated stage cost function ℓ^{int}, all states in the RCI set Ω are weighted equally. For many examples, this can provide better results than using min-max approaches, where only the worst case is considered. However, the performance bound in (4.7) is provided only with respect to the closed-loop nominal sequence and the integrated stage cost function. Hence, in contrast to min-max robust economic MPC based on tubes (see, e.g., the performance statement in (3.29)), we can only derive statements on the averaged performance.*

4.2 Optimal steady-state operation

In this section, we focus on optimal operation at steady-state for systems affected by disturbances in the presented tube-based robust economic MPC scheme. First, we investigate optimality for the approach based on Problem 4.1, where we have to take into account that the nominal initial state is an optimization variable. In a second step, the nominal initial state is fixed (see Remarks 2.18 and 3.10) and optimality is provided for this simpler approach, which follows directly from the nominal approach in Angeli et al. (2012).

Before being able to investigate optimality for the tube-based robust economic MPC approach presented in Algorithm 4.2, one has to discuss the behavior of the closed-loop nominal state sequence z^*. Since the nominal initial state is an optimization variable, we cannot consider a trajectory of the nominal dynamics (2.12), but we have to investigate the possible evolution of z^*.

The connection of the nominal initial states at two consecutive time instances, that is, $z(0|t)$ and $z(0|t+1)$, is depicted in Figure 4.1. Given $z(0|t)$ and the RCI set Ω, we know (by design) that $x(t) \in \{z(0|t)\} \oplus \Omega$ and due to the constraints (4.4) that $x(t+1) \in \{z(1|t)\} \oplus \Omega$. However, $z(0|t+1)$ is unknown at time t. By using

$$\overline{\mathbb{X}}_{\text{init}}(z) := \left\{ \bar{z} \in \overline{\mathbb{X}} : \exists x \in \{z\} \oplus \Omega, x - \bar{z} \in \Omega \right\}, \tag{4.8}$$

the best guess at time t knowing only $z(0|t)$ and $v(0|t)$ is $z(0|t+1) \in \overline{\mathbb{X}}_{\text{init}}(f(z(0|t), v(0|t), 0))$. We note that $\overline{\mathbb{X}}_{\text{init}}(z) = (\{z\} \oplus \Omega \oplus (-\Omega)) \cap \overline{\mathbb{X}}$. Thus, for RCI sets which are symmetric with respect to the origin, it holds that $\overline{\mathbb{X}}_{\text{init}}(z) = (\{z\} \oplus 2\Omega) \cap \overline{\mathbb{X}}$. With this set, it follows for the closed-loop nominal sequence z^* resulting from Algorithm 4.2 that $z^*(0|t+1) \in \overline{\mathbb{X}}_{\text{init}}(f(z^*(0|t), v^*(0|t), 0))$.

Next, we define optimal operation at steady-state in the tube-based setting (compare to Definition 2.24):

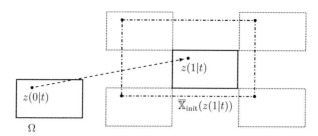

Figure 4.1: Given the sets $\{z(0|t)\} \oplus \Omega$ (solid left), the real system state $x(t+1)$ is guaranteed to satisfy $x(t+1) \in \{z(1|t)\} \oplus \Omega$ (solid right). Thus, *all* possible nominal initial states $z(0|t+1)$ satisfying $x(t+1) \in \{z(0|t+1)\} \oplus \Omega$ for all possible $x(t+1) \in \{z(1|t)\} \oplus \Omega$ are in $\overline{\mathbb{X}}_{\text{init}}(z(1|t))$ (dash-dotted). The dotted sets represent extreme cases where $x(t+1)$ lies on the boundary of $\{z(1|t)\} \oplus \Omega$.

Definition 4.5 (Ω-robust optimal operation at steady-state). *System* (4.1) *is* Ω-*robustly optimally operated at steady-state* with respect to the cost function ℓ and the constraints (2.2), *if for each pair of state and input sequence satisfying* $(z(k), v(k)) \in \overline{\mathbb{Z}}$ *and* $z(k+1) \in \overline{\mathbb{X}}_{\text{init}}(f(z(k), v(k), 0))$, *it holds that*

$$\liminf_{T \to \infty} \frac{1}{T} \sum_{t=0}^{T-1} \ell^{\text{int}}\Big(z(k), v(k)\Big) \geq \ell^{\text{int}}(z_s^{\text{int}}, v_s^{\text{int}}), \tag{4.9}$$

where $(z_s^{\text{int}}, v_s^{\text{int}})$ *is the robust optimal steady-state according to* (4.6).

According to this definition, a system is Ω-robustly optimally operated at steady-state, if staying with the nominal sequence at the robust optimal steady-state provides the best average integral performance compared to the average integral performance along any other feasible nominal closed-loop sequence. Thus, this definition also holds for the real system (4.1), averaged over all possible disturbances by integrating over the RCI set Ω. This is a quite conservative definition, since the underlying sequence for the nominal states considers an over-approximation of the actual closed-loop sequence to be achieved. However, this statement is made a priori, and thus, we cannot incorporate more information.

As discussed in Section 2.3, in the nominal case optimality is directly related to the concept of dissipativity. Hence, with the new notion of optimality provided in Definition 4.5, we must also adapt the idea of dissipativity to the robust setup considered:

Definition 4.6 (Ω-robust dissipativity). *System* (4.1) *is* Ω-*robustly dissipative* on a set $\mathbb{S} \subseteq \overline{\mathbb{Z}}$ *with respect to the supply rate* $s : \overline{\mathbb{Z}} \to \mathbb{R}$ *if there exists a storage function* $\lambda : \mathbb{S}_{\overline{\mathbb{X}}} \to \mathbb{R}_{\geq 0}$ *such that the following inequality is satisfied for all* $(z, v) \in \mathbb{S}$:

$$\sup_{x \in \overline{\mathbb{X}}_{\text{init}}(f(z,v,0))} \lambda(x) - \lambda(z) \leq s(z, v). \tag{4.10}$$

We can define the largest set of all state and input pairs in $\overline{\mathbb{Z}}$ which are part of a feasible nominal sequence by

$$\overline{\mathbb{Z}}_{\text{init}}^o := \Big\{(z, v) \in \overline{\mathbb{Z}} : \exists \, \tilde{\boldsymbol{v}} \text{ s.t. } \big(\tilde{z}(0), \tilde{v}(0)\big) = (z, v),$$
$$\tilde{z}(k+1) \in \overline{\mathbb{X}}_{\text{init}}\Big(f(\tilde{z}(k), \tilde{v}(k), 0)\Big), \big(\tilde{z}(k), \tilde{v}(k)\big) \in \overline{\mathbb{Z}}, \forall k \in \mathbb{I}_{\geq 0}\Big\}. \tag{4.11}$$

The considered sequences in $\overline{\mathbb{Z}}^o_{\text{init}}$ need *not* satisfy the nominal dynamics (2.12), which is in accordance to the sequence of nominal initial states resulting from applying Algorithm 4.2, i.e., $\mathbf{z}^*$.

With these definitions, we can derive the main result of this section:

Theorem 4.7. *Consider system (4.1) and let Assumptions 2.7–2.9 be satisfied. Suppose that system (4.1) is Ω-robustly dissipative on $\overline{\mathbb{Z}}^o_{\text{init}}$ with respect to the supply rate*

$$s(z,v) = \ell^{\text{int}}(z,v) - \ell^{\text{int}}(z^{\text{int}}_s, v^{\text{int}}_s), \tag{4.12}$$

then the system is Ω-robustly optimally operated at steady-state.

We note that we have to consider Ω-robust dissipativity, since no other connection of $z(0|t)$ and the consecutive nominal state $z(0|t+1)$ is provided.

4.2.1 Fixing nominal initial state

In Remark 2.18 and 3.10, we have discussed to fix the nominal dynamics by replacing (4.4b) in Problem 4.1 by

$$z(0|t+1) = f(z(0|t), v^*(0|t), 0). \tag{4.13}$$

Considering $z(0|0)$, one can choose any state satisfying $x_0 \in \{z(0|0)\} \oplus \Omega$, e.g., $z(0|0) = x_0$. By this new constraint, the nominal initial state $z(0|t)$ is no longer an optimization variable, but determined by the solution to the optimization problem at time $t-1$. Moreover by replacing (4.4b) with (4.13), the modified problem is independent of $x(t)$ (except for $z(0|0)$). With this new constraint, Algorithm 4.2 must be adopted respectively:

Algorithm 4.8 (Tube-based robust economic model predictive control with fixed nominal initial state). *At each time instant $t \in \mathbb{I}_{\geq 0}$, measure the state $x(t)$ and solve Problem 4.1 with (4.4b) replaced by (4.13). Apply the control input $u(t) := \varphi\big(v^*(0|t), x(t), z^*(0|t)\big)$ to system (4.1), and $v(t) = v^*(0|t)$ to the nominal system (2.12).*

The additional constraint, which limits the closed-loop nominal sequence to a trajectory of the nominal system (2.12), does not influence the performance statement in Theorem 4.3 but allows for a simpler dissipativity statement in order to provide optimal operation at steady-state. Hence, one can restate Definition 4.5 such that it is limited to feasible state and input trajectories (providing *robust optimal operation at steady-state*[1]). Within Definition 4.6, we can cope with the limitation by replacing (4.10) with

$$\lambda\big(z(0|t+1)\big) - \lambda\big(z(0|t)\big) \leq s\big(z(0|t), v(0|t)\big), \tag{4.14}$$

that is, with dissipativity for the nominal dynamics, since it holds that $z(0|t+1) = z(1|t) = f(z(0|t), v(0|t), 0)$ by Algorithm 4.8. In the following discussion, we refer to this as *robust dissipativity*. Similar to Theorem 4.7 and exactly as in Angeli et al. (2012, Proposition 6.4), one can derive that robust dissipativity leads to robust optimal operation at steady-state.

We note that robust dissipativity is usually a statement simpler to proof as well as to satisfy than Ω-robust dissipativity. From a conceptual point of view, Ω-robust dissipativity

---

[1]System (4.1) is *robustly optimally operated at steady-state* if (4.9) is satisfied for each solution of the nominal dynamics (2.12) satisfying $(z(k), v(k)) \in \overline{\mathbb{Z}}$ for all $k \in \mathbb{I}_{\geq 0}$.

is more restrictive than robust dissipativity. This is since Ω-robust dissipativity has to hold for all possible sequences $z^*(0|\cdot)$ resulting from applying Algorithm 4.2 which need not be solutions of the nominal system (2.1). In particular, every system that is Ω-robustly dissipative must be robustly dissipative, but not vice versa.

For nominal systems, it is well known that linear systems with an arbitrary convex stage cost function are optimally operated at steady-state (see, e.g., Angeli et al. (2009); Diehl et al. (2011)). Thus, we investigate if this holds also for the robust counterpart:

Proposition 4.9. *Consider the linear system (2.20). Let Assumptions 2.2, 2.3, and 2.7–2.10 be satisfied and let the input be parametrized according to (2.21). Moreover, let ℓ and $\mathbb{Z}$ be convex. Then the system is robustly optimally operated at steady-state.*

If ℓ^{int} preserves convexity, the proof follows directly from Angeli et al. (2009, Theorem 4). Thus, we consider the following lemma providing also the proof for Proposition 4.9:

Lemma 4.10. *Consider the linear system (2.20). Let Assumptions 2.2, 2.3, and 2.7–2.9 be satisfied and let the input be parametrized according to (2.21). Moreover, let ℓ and $\mathbb{Z}$ be convex and let Ω be an RPI set for the error dynamics (2.22). Then the integrated stage cost $\ell^{\mathrm{int}}(z,v)$ given in (4.3) is convex.*

Remark 4.11. *Proposition 4.9 only holds if the closed-loop nominal sequence is a trajectory of the nominal system (2.12), that is, it satisfies (4.13). If the nominal initial state $z(0|t)$ is an optimization variable, convexity of the stage cost is not sufficient for Ω-robust optimal operation at steady-state.*

4.2.2 Adding dissipativity-inducing constraint

The first approach to circumvent the restrictive notion of Ω-robust dissipativity was to fix the free nominal initial state to a particular value. However, this might be conservative, especially when taking the closed-loop average performance into account.

While considering Ω-robust dissipativity (4.10) allows for any feasible nominal initial state, robust dissipativity (4.14) only takes one particular state into account. Thus, we now provide an approach that allows for optimizing over the nominal initial state while requiring only robust dissipativity.

When only assuming robust dissipativity while keeping the nominal initial state as a free variable in the optimization, the proof of Theorem 4.7 does no longer hold. To overcome this problem, we need to provide that the nominal initial state has a certain relation to the nominal trajectory at the previous step with respect to the value of the storage function. Taking a closer look at the second inequality in (A.9), it is sufficient to guarantee (in addition to robust dissipativity) that

$$\lambda\big(z(0|t)\big) \leq \lambda\big(z^*(1|t-1)\big), \tag{4.15}$$

meaning that the storage at the next nominal initial state may not be greater than the storage at the first predicted nominal state of the previous iterate. This additional condition induces Ω-robust optimal operation at steady-state while still allowing for more nominal initial states than (4.13).

However, this constraint must be added to Problem 4.1 leading to more constraints in the optimization problem. Moreover, the storage function λ need usually *not* be known when implementing the algorithm and is only needed in the proof of optimal operation. In this setup, one would need to determine λ to be able to implement the setup.

Remark 4.12. *The condition stated above in (4.15) can even be relaxed further by considering*

$$\lambda\big(z(0|t)\big) \leq s\big(z^*(0|t-1), v^*(0|t-1)\big) + \lambda\big(z^*(0|t-1)\big) \tag{4.16}$$

as an additional constraint within the optimization problem, where $z^*(0|t-1)$ and $v^*(0|t-1)$ are the nominal state and input determined (and applied) at the previous iteration. If the system is robustly dissipative, a feasible $z(0|t)$ always exists. When taking the definition of Ω-robust dissipativity into account, one can see that all Ω-robustly dissipative systems satisfy condition (4.16) for all $z(0|t) \in \overline{\mathbb{X}}_{\text{init}}(z^*(1|t-1))$. This shows again the conservatism of Ω-robust dissipativity, since one cannot—in contrast to the constraint (4.16) to be appended—take information about the "current" nominal state into account.

4.2.3 Converse results (necessity of dissipativity)

As introduced in Section 2.3, a converse dissipativity statement can be investigated for nominal economic MPC. The necessity of dissipativity is based on specific controllability assumptions. In this section, we investigate a similar statement for the disturbed case.

Similar to Müller et al. (2015b), the following set definitions provide a controllability/reachability assumption. In contrast to Müller et al. (2015b), those sets must account for the changes in the nominal initial state, again by consideration of $\overline{\mathbb{X}}_{\text{init}}$. We introduce the M-step controllable set, that is, the set of all states for with the robust optimal steady-state can be reached in M steps, by

$$\overline{\mathcal{C}}_{M,\text{init}} := \Big\{ x \in \mathbb{R}^n : \exists \boldsymbol{v} \text{ s.t. } z(0) = x, z(k+1) \in \overline{\mathbb{X}}_{\text{init}}\big(f(z(k), v(k), 0)\big), \\ z_s^{\text{int}} \in \overline{\mathbb{X}}_{\text{init}}\big(f(z(M-1), v(M-1), 0)\big), \big(z(k), v(k)\big) \in \overline{\mathbb{Z}}, \ \forall k \in \mathbb{I}_{[0,M-1]} \Big\}, \tag{4.17}$$

(compare to (2.40)) and the set of states reachable from the robust optimal steady-state in M steps by

$$\overline{\mathcal{R}}_{M,\text{init}} := \Big\{ x \in \mathbb{R}^n : \exists \boldsymbol{v} \text{ s.t. } z(0) = z_s^{\text{int}}, z(k+1) \in \overline{\mathbb{X}}_{\text{init}}\big(f(z(k), v(k), 0)\big), \\ z(M) = x, \big(z(k), v(k)\big) \in \overline{\mathbb{Z}}, \ \forall k \in \mathbb{I}_{[0,M]} \Big\} \tag{4.18}$$

(see (2.41)). We define the set of nominal state and input pairs which are part of a feasible state/input sequence pair $(\boldsymbol{z}, \boldsymbol{u})$ guaranteeing the state sequence to stay in the intersection of $\overline{\mathcal{C}}_{M,\text{init}}$ and $\overline{\mathcal{R}}_{M,\text{init}}$ (as in (2.42)) by

$$\overline{\mathcal{Z}}_{M,\text{init}} := \Big\{ (x, u) \in \overline{\mathbb{Z}} : \exists \boldsymbol{v} \text{ s.t. } \big(z(0), v(0)\big) = (x, u), z(k+1) \in \overline{\mathbb{X}}_{\text{init}}\big(f(z(k), v(k), 0)\big), \\ \big(z(k), v(k)\big) \in \overline{\mathbb{Z}}, z(k) \in \overline{\mathcal{C}}_{M,\text{init}} \cap \overline{\mathcal{R}}_{M,\text{init}}, \ \forall k \in \mathbb{I}_{\geq 0} \Big\}. \tag{4.19}$$

With these sets defined, we can state the main result of this section:

Theorem 4.13. *Suppose that system (4.1) is Ω-robustly optimally operated at steady-state and let Assumptions 2.1, 2.3, and 2.7–2.9 be satisfied. Then, for each $M \in \mathbb{I}_{\geq 1}$, system (4.1) is Ω-robustly dissipative on $\overline{\mathcal{Z}}_{M,\text{init}}$ with respect to the supply rate (4.12).*

We note that $\overline{\mathcal{Z}}_{M,\text{init}} \subseteq \overline{\mathbb{Z}}^o_{\text{init}}$. If $\overline{\mathcal{Z}}_{M,\text{init}} = \overline{\mathbb{Z}}^o_{\text{init}}$ for some $M \in \mathbb{I}_{\geq 1}$, then we have an if-and-only-if relation of Ω-robust optimal operation at steady-state and Ω-robust dissipativity on $\overline{\mathbb{Z}}^o_{\text{init}}$ for the respective system (compare to Müller et al. (2015b, Corollary 1)). As discussed for the nominal case in Section 2.3.1, the converse result (under the mild controllability assumption) provides that Ω-robust dissipativity is not too conservative for characterizing Ω-robust optimal operation at steady-state.

The proof of Theorem 4.13 follows along the lines of Müller et al. (2015b), and subsequently, we only present the major steps: The proof follows by contradiction. One assumes that the system is Ω-robustly optimally operated at steady-state, that is, it satisfies (4.9) for all possible closed-loop sequences, but is *not* Ω-robustly dissipative. Following the results stated in Willems (1972), the available storage

$$S_a(x) := \sup_{\substack{T \geq 0 \\ z(0)=x,\, z(k+1)\in\overline{\mathbb{X}}_{\text{init}}(f(z(k),v(k),0)), \\ (z(k),v(k))\in\overline{\mathbb{Z}}^o_{\text{init}},\forall k\in\mathbb{I}_{\geq 0}}} \sum_{k=0}^{T-1} -s\Big(z(k),v(k)\Big), \qquad (4.20)$$

satisfies $S_a(x) < \infty$ for all $x \in \overline{\mathbb{X}}^o_{\text{init}}$ if and only if system (4.1) is Ω-robustly dissipative. From the assumption that the system is not Ω-robustly dissipative, one constructs a cyclic state/input sequence pair (z, v), such that these sequences violate (4.9). This provides the desired contradiction.

Some comments regarding the available storage (4.20) are in order. We highlight that the available storage considered in this setup is, again, based on $\overline{\mathbb{X}}_{\text{init}}$, that is, we do not consider nominal trajectories but sequences that take the possibility of a free nominal initial state into account. Moreover, following the arguments in Willems (1972), one can show that the available storage (4.20) can be employed as a storage function to show Ω-robust dissipativity.

Remark 4.14. *We have only shown the converse statement for tube-based robust economic MPC with respect to Algorithm 4.2, that is, where the nominal initial state is an optimization variable. In case the nominal initial state is fixed according to (4.13), the converse statement follows directly from Müller et al. (2015b, Theorem 3), replacing the tube-dynamics by the nominal dynamics (2.12), both in the controllability and reachability sets ($\mathcal{C}_M$, $\mathcal{R}_M$, and $\mathcal{Z}_M$ in (2.40), (2.41), and (2.42), respectively) as well as in the definition of the available storage.*

4.2.4 Connection to min-max robust economic MPC

Due to its structural similarity, the min-max robust economic MPC based on tubes presented in Section 3.2 can also be investigated for optimal operating regimes by means of the statements presented above. Comparing Problem 3.6 with Problem 4.1, the problems differ in two aspects: First, a different input parametrization is used. As discussed in Remark 3.4 they are not equivalent. However, the input parametrization (2.14) can also be employed in the min-max robust economic MPC approach based on tubes. In Chapter 3, this was avoided to keep the line of presentation. The second difference is the stage cost function considered in the objectives (3.18) and (4.5), respectively. These represent the way of treating the disturbance within the objective. Replacing the input parametrization in (3.18) by the parametrization in (2.14), we receive

$$\ell^{\max}(z, v) = \max_{\omega \in \Omega} \ell\Big(z + \omega, \varphi(v, z + \omega, z)\Big). \tag{4.21}$$

Thus, one can see that all results regarding optimal operating regimes derived for the tube-based robust economic MPC also hold for the min-max robust economic MPC based on tubes by replacing ℓ^{int} with $\ell^{\max}$ and using the associated robust optimal steady-state (3.19). In particular, one can show that Theorem 4.7 and Theorem 4.13 hold for the min-max robust economic MPC setup based on tubes. Proposition 4.9 holds true for $\ell^{\max}$ as well, the proof for preserving convexity under the maximization with ℓ given convex is provided in Boyd and Vandenberghe (2004, Section 3.2.3).

Even though the two setups—tube-based robust economic MPC and min-max robust economic MPC based on tubes—have the same definitions regarding dissipativity, the statements are not equivalent in the sense that a system which is Ω-robustly dissipative with respect to the integrated cost need not be Ω-robustly dissipative with respect to the maximized cost and vice versa.

Example 4.15. *We consider a linear scalar system $x(t + 1) = 0.5x(t) + u(t) + w(t)$, with $\mathbb{X} = \{x \in \mathbb{R} : |x| \leq 5\}$, $\mathbb{U} = \{u \in \mathbb{R} : |u| \leq 2.3\}$, $\mathbb{W} = \{w \in \mathbb{R} : |w| \leq 0.1\}$, and $\varphi(v, x, z) = v$. The mRPI set can be computed as $\Omega = [-0.2, 0.2]$. The stage cost is given by*

$$\ell(x, u) = \begin{cases} -(x - 4.6)^2 & \text{for } 4.6 \leq x, \\ 0 & \text{else.} \end{cases}$$

Tightening the constraint set leads to $\overline{\mathbb{Z}} = ((\mathbb{X} \ominus \Omega) \times \mathbb{U}) = ([-4.8, 4.8] \times [-2.3, 2.3])$. The integrated stage cost is

$$\ell^{\mathrm{int}}(z, v) = \begin{cases} -\frac{2}{5}z^2 + \frac{92}{25}z - \frac{3176}{375} & \text{for } 4.8 \leq z, \\ -\frac{1}{3}(z - 4.4)^3 & \text{for } 4.4 \leq z < 4.8, \\ 0 & \text{else,} \end{cases}$$

and the maximized stage cost

$$\ell^{\max}(z, v) = \begin{cases} -(z - 4.8)^2 & \text{for } 4.8 \leq z, \\ 0 & \text{else.} \end{cases}$$

Computing the associated robust optimal stead-states, it follows that $(z_s^{\mathrm{int}}, v_s^{\mathrm{int}}) = (4.6, 2.3)$ and $(z_s^{\max}, v_s^{\max}) = [(-4.6, -2.3), (4.6, 2.3)]$, meaning we have infinitely many robust optimal steady-states in the min-max case. For the min-max case, all states within $\overline{\mathbb{Z}}$ have the same cost. Thus when investigating Ω-robust dissipativity,

$$\sup_{x \in (\{0.5z + v\} \oplus 2\Omega_\infty) \cap \overline{\mathbb{X}}} \lambda^{\max}(x) - \lambda^{\max}(z) \leq 0$$

is trivially satisfied by $\lambda^{\max}(z) = 0$. For the integrated cost, we have to prove

$$\sup_{x \in (\{0.5z + v\} \oplus 2\Omega_\infty) \cap \overline{\mathbb{X}}} \lambda^{\mathrm{int}}(x) - \lambda^{\mathrm{int}}(z) \leq \ell^{\mathrm{int}}(z, v) + 0.002667$$

for all $(z, u) \in \overline{\mathbb{Z}}^o_{\mathrm{init}}$. *The most critical state is* $(z, v) = (4.8, 2.3)$. *One can see that this state is in* $\overline{\mathbb{Z}}^o_{\mathrm{init}}$ *since* $z^+ = 4.7$. *Hence, we have to investigate if there exists some non-negative function* λ^{int} *such that*

$$\sup_{x \in (\{4.7\} \oplus 2\Omega_\infty) \cap \overline{\mathbb{X}}} \lambda^{\mathrm{int}}(x) - \lambda^{\mathrm{int}}(4.8) \leq -0.0187.$$

Since $4.8 \in (\{4.7\} \oplus 2\Omega_\infty) \cap \overline{\mathbb{X}}$, *the left hand side is non-negative independent of the choice of* λ^{int}, *and hence, it is impossible to find a storage function satisfying this condition.*

4.3 Asymptotic stability

As discussed in Chapter 2, a particular feature of economic MPC is that the resulting closed-loop system does not necessarily exhibit convergent behavior. However, it was shown in the previous section that a certain dissipativity condition implies optimal operation of the system at steady-state. In this case, one would also like (a neighborhood of) the robust optimal steady-state to be stable for the closed-loop system. In the nominal case, we have discussed in Theorem 2.28 that this can be the case. We show that a similar result can be established also in our setting with disturbances.

To this end, we consider a more restrictive dissipativity statement, that is, *strict Ω-robust dissipativity*:

Definition 4.16 (Strict Ω-robust dissipativity). *System* (4.1) *is strictly Ω-robustly dissipative on a set* $\mathbb{S} \subseteq \overline{\mathbb{Z}}$ *with respect to the supply rate* $s : \overline{\mathbb{Z}} \to \mathbb{R}$ *if there exists a continuous storage function* $\lambda : \mathbb{S}_{\overline{\mathbb{X}}} \to \mathbb{R}$ *and a positive definite function* $\rho : \mathbb{S}_{\overline{\mathbb{X}}} \to \mathbb{R}_{\geq 0}$ *such that the following inequality is satisfied for all* $(z, v) \in \mathbb{S}$:

$$\max_{x \in \overline{\mathbb{X}}_{\mathrm{init}}(f(z,v,0))} \lambda(x) - \lambda(z) \leq s(z, v) - \rho(z - z_s^{\mathrm{int}}). \tag{4.22}$$

In contrast to the definition of Ω-robust dissipativity (see Definition 4.6), we require the storage function to be continuous instead of being merely bounded from below. This is due to the subsequent proof for asymptotic stability where continuity of the storage function is required. Moreover, we can replace the supremum in Definition 4.6 by the maximum, since we consider continuous functions over compact sets (consider Assumption 2.8).

Based on strict Ω-robust dissipativity, we can state the following result on asymptotic stability:

Theorem 4.17. *Consider system* (4.1) *and let* Ω *be the RCI set for the associated error dynamics* (2.15). *Moreover, let Assumptions 2.1, 2.3, 2.7–2.9, and 4.1 be satisfied and let the system be strictly Ω-robustly dissipative with respect to the supply rate* (4.12). *Then, the closed loop resulting from applying Algorithm 4.2 asymptotically converges to the set* $\{z_s^{\mathrm{int}}\} \oplus \Omega$. *The region of attraction is* $\overline{\mathbb{X}}_N \oplus \Omega$.

Furthermore, if the storage function λ *in addition satisfies for all* $x \in (\overline{\mathbb{X}}_{\mathrm{init}}(z_s^{\mathrm{int}}) \oplus \overline{\mathcal{B}}_\epsilon) \cap \overline{\mathbb{X}}$

$$\lambda(x) - \lambda(z_s^{\mathrm{int}}) \leq 0, \tag{4.23}$$

with some $\epsilon > 0$ *such that* $\{z_s^{\mathrm{int}}\} \oplus \overline{\mathcal{B}}_\epsilon \subseteq \overline{\mathbb{X}}_\mathrm{f}$, *then the set* $\{z_s^{\mathrm{int}}\} \oplus \Omega$ *is asymptotically stable for the closed loop resulting from applying Algorithm 4.2 with region of attraction* $\overline{\mathbb{X}}_N \oplus \Omega$.

Remark 4.18. *Considering strict Ω-robust dissipativity as required in Theorem 4.17, one can see that the storage function must satisfy $\lambda(z) \leq \lambda(z_s^{\mathrm{int}})$ for all $z \in \overline{\mathbb{X}}_{\mathrm{init}}(z_s^{\mathrm{int}})$. This follows from $\max_{x \in \overline{\mathbb{X}}_{\mathrm{init}}(z_s^{\mathrm{int}})} \lambda(x) - \lambda(z_s^{\mathrm{int}}) \leq 0$ and from continuity of λ. In fact, this can only be guaranteed if the integrated stage cost function ℓ^{int} is positive definite with respect to z_s^{int} and v_s^{int} in a neighborhood of z_s^{int} and v_s^{int}.*

However, as we have seen in the proof, strict Ω-robust dissipativity is not enough in order to ensure asymptotic stability, since this allows to draw conclusions about the behavior of the storage function only on $\overline{\mathbb{X}}_{\mathrm{init}}(z_s^{\mathrm{int}})$. In order to be able to guarantee stability, we have to consider system states in the neighborhood of $\{z_s^{\mathrm{int}}\} \oplus \Omega$. Hence, by considering the nominal initial state as an optimization variable, we also have to take states in the neighborhood of $\overline{\mathbb{X}}_{\mathrm{init}}(z_s^{\mathrm{int}})$ into account. Thus, as follows in the proof, we must extend the neighborhood of z_s^{int} and v_s^{int} in which the integrated stage cost is positive definite. This is achieved by (4.23).

Remark 4.19. *The results for asymptotic stability also apply for the min-max robust economic MPC approach based on tubes when employing the respective cost function (ℓ^{max} instead of ℓ^{int}) and robust optimal steady-state (($z_s^{\mathrm{max}}, v_s^{\mathrm{max}}$) instead of ($z_s^{\mathrm{int}}, v_s^{\mathrm{int}}$)) (compare to Section 4.2.4).*

4.3.1 Fixing nominal initial state—revisited for stability

As in the discussion on optimal steady-state operation in Section 4.2.1, we investigate stability when fixing the nominal initial state to satisfy the nominal dynamics by replacing (4.4b) with (4.13). As mentioned before, this causes the optimization problem to be independent of the currently measured state $x(t)$. Thus, we expect a slightly weaker statement considering stability. Taking Algorithm 4.8 into account, we investigate the closed-loop composite system

$$z(t+1) = f\Big(z(t), v^*(0|t), 0\Big), \qquad\qquad z(0) = z_0, \qquad (4.24a)$$

$$x(t+1) = f\Big(x(t), \varphi\big(v^*(0|t), x(t), z(t)\big), w(t)\Big), \qquad x(0) = x_0. \qquad (4.24b)$$

With respect to this composite system, the following statement can be derived which is a counterpart of Rawlings and Mayne (2009, Proposition 3.15) in the framework of robust economic MPC:

Theorem 4.20. *Consider system (4.1) and let Ω be the RCI set for the associated error dynamics (2.15). Moreover, let Assumptions 2.1, 2.3, 2.7–2.9, and 4.1 be satisfied and let system (4.1) be strictly robustly dissipative[2] with respect to the supply rate (4.12). Then, the set $\mathcal{A} := \{z_s^{\mathrm{int}}\} \times (\{z_s^{\mathrm{int}}\} \oplus \Omega)$ is asymptotically stable for the closed-loop composite system (4.24) resulting from applying Algorithm 4.8 with region of attraction $\overline{\mathbb{X}}_N \times (\overline{\mathbb{X}}_N \oplus \Omega)$.*

The statement that $\mathcal{A}$ is asymptotically stable for the closed-loop composite system is weaker than $\{z_s^{\mathrm{int}}\} \oplus \Omega$ being asymptotically stable for the closed-loop real system as provided in Theorem 4.17. In contrast, the required assumption reduces to strict robust dissipativity, which is in general less restrictive than Ω-robust strict dissipativity.

---

[2]System (4.1) is *strictly robustly dissipative* on a set $\mathbb{S} \subseteq \overline{\mathbb{Z}}$ with respect to the supply rate $s : \overline{\mathbb{Z}} \to \mathbb{R}$ if there exists a continuous storage function $\lambda : \mathbb{S}_{\overline{\mathbb{X}}} \to \mathbb{R}$ and a positive definite function $\rho : \mathbb{S}_{\overline{\mathbb{X}}} \to \mathbb{R}_{\geq 0}$ such that the following inequality is satisfied for all $(z,v) \in \mathbb{S}$: $\lambda(f(z,v,0)) - \lambda(z) \leq s(z,v) - \rho(z - z_s^{\mathrm{int}})$.

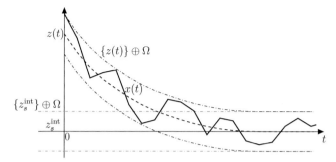

Figure 4.2: Exemplary trajectories of the nominal state $z(t)$ (dashed) and the real state $x(t)$ (solid). The set $\mathcal{A}$ is asymptotically stable for the composite system (4.24) and, thus, the robust optimal steady state z_s^{int} is asymptotically stable for the nominal system (4.24a). However, the set $\{z_s^{\text{int}}\} \oplus \Omega$ (gray dash-dotted) is not asymptotically stable for real system (4.24b) since the state trajectory of the real system reaches this set, but leaves it again.

Remark 4.21. *As discussed in Section 4.2.1, we know that for feasibility $x_0 \in \{z_0\} \oplus \Omega$ must be satisfied with respect to the nominal initial state z_0. Then again, the nominal initial state can be used to provide that $\{z_s^{\text{int}}\} \oplus \Omega$ is asymptotically stable for the real closed-loop system. Taking the proof of Theorem 4.20 into account, this can be achieved by relating $|z_0 - z_s^{\text{int}}|$ with $|x_0|_{\{z_s^{\text{int}}\} \oplus \Omega}$ by comparison functions. However, due to the structure of Algorithm 4.8, the nominal initial state is only free for the very first iteration, that is at $t = 0$. For all $t \in \mathbb{I}_{\geq 1}$, the nominal state is determined by the nominal dynamics. Thus, even if one can relate $|z_0 - z_s^{\text{int}}|$ with $|x_0|_{\{z_s^{\text{int}}\} \oplus \Omega}$, one does not achieve uniform asymptotic stability, but the weaker notion of horizon-∞-asymptotic stability (Brunner and Allgöwer, 2015). The problem for this setup becomes apparent considering the case in Figure 4.2: The system is close to $\{z_s^{\text{int}}\} \oplus \Omega$, that is, close to the set to be stabilized. Now a "beneficial" disturbance pushes the state of the real system into the set $\{z_s^{\text{int}}\} \oplus \Omega$. Since the nominal state is only determined with respect to the very initial real state, the algorithm cannot react to this disturbance and will possibly force the real system to leave the set $\{z_s^{\text{int}}\} \oplus \Omega$ again. Only eventually will the state of the real system converge. Thus, we do not achieve uniform asymptotic stability.*

4.3.2 Adding dissipativity-inducing constraint—revisited for stability

As presented in Section 4.2.2, satisfying the strong assumption of Ω-robust dissipativity can be circumvented either by fixing the nominal initial state to a trajectory of the nominal system (as provided in the previous section) or by means of an additional constraint on the nominal initial state. Under strict robust dissipativity, one can employ the constraint provided in (4.15) in order to show asymptotic convergence. Taking also the additional assumption on the storage function in (4.23) into account, asymptotic stability of $\{z_s^{\text{int}}\} \oplus \Omega$ follows for the closed-loop system along the same lines as in Theorem 4.17.

Regarding the constraint on the nominal initial states in Remark 4.12, one can make use of an equivalent constraint, however, strict dissipativity must be considered. This leads to a different constraint to be added to Problem 4.1, namely

$$\lambda\Big(z(0|t)\Big) \leq s\Big(z^*(0|t-1), v^*(0|t-1)\Big) + \lambda\Big(z^*(0|t-1)\Big) - \rho\Big(z^*(0|t-1) - z_s^{\text{int}}\Big), \quad (4.25)$$

where again $z^*(0|t-1)$ and $v^*(0|t-1)$ are the nominal state and input determined (and applied) at the previous iteration.

Of course, the same drawbacks as in Section 4.2.2 arise. We need to modify Problem 4.1 by an additional, possibly nonlinear constraint. Moreover, an appropriate storage function λ and positive definite function ρ allowing to provide strict robust dissipativity are required.

4.4 Average constraints

In the following section, we impose additional average constraints on the system. In contrast to pointwise-in-time constraints as in (2.2), these constraints are required to be satisfied on average only and not at each time instant $t \in \mathbb{I}_{\geq 0}$. Due to the possibly non-convergent closed-loop behavior, constraints on the averages of state and input variables become of interest in economic MPC. These constraints have been investigated for the undisturbed setup in (Angeli et al., 2011, 2012; Müller et al., 2013; Müller et al., 2014b). Interestingly, these constraints require to be dealt with online. This is in contrast to stabilizing MPC, where asymptotic averages of state and input variables are determined by their value at the set-point to be stabilized, and hence, asymptotic average constraints can be considered offline (when determining the set-point). The concept of average constraints can be interesting for different applications such as a chemical reactor, where the average feed flow should be constrained in order to meet storage capacities, or a supply chain network, where the average inflow should be constrained. The results in this section are based on (Bayer and Allgöwer, 2014) and (Bayer et al., 2015b).

In order to handle asymptotic average constraints, we consider the auxiliary output

$$y(t) = h\Big(x(t), u(t)\Big), \quad (4.26)$$

where $h : \mathbb{R}^n \times \mathbb{R}^m \to \mathbb{R}^p$ and $y(t) \in \mathbb{R}^p$ is the output at time t.

Assumption 4.2. *The auxiliary output function h is continuous.*

With this auxiliary output, the average constraints imposed on system (4.1) are given by

$$\text{Av}\Big[h(\boldsymbol{x}, \boldsymbol{u})\Big] \subseteq \mathbb{R}_{\leq 0}^p. \quad (4.27)$$

Stating the average constraint with respect to the negative orthant is not a major restriction, since h can be an arbitrary nonlinear function (see, e.g., Angeli et al. (2012); Müller et al. (2014b) where the average constraints are provided for some closed and convex set).

For the *undisturbed* system, satisfaction of the asymptotic average constraint can be guaranteed by imposing the additional constraint, see, e.g., Angeli et al. (2012); Müller et al. (2014b),

$$\sum_{k=0}^{N-1} h\Big(x(k|t), u(k|t)\Big) \in \mathbb{Y}_t, \quad (4.28)$$

where the time-varying set $\mathbb{Y}_t$ is provided by the recursive formula

$$\mathbb{Y}_{t+1} = \mathbb{Y}_t \oplus \mathbb{R}^p_{\leq 0} \oplus \left\{ -h(x(t), u(0|t)) \right\}, \qquad \mathbb{Y}_0 = \mathbb{Y}_{00} \oplus \mathbb{R}^p_{\leq 0}. \tag{4.29}$$

The set $\mathbb{Y}_{00}$ is an arbitrary compact set containing the optimal steady-state.

As in the previous sections, we cannot predict the exact system behavior due to the disturbances, but we only know bounds on the behavior of the real system by means of the RCI set Ω. Hence, in order to guarantee satisfaction of the average constraint (4.27), one would need to add the following constraint to Problem 4.1:

$$\sum_{k=0}^{N-1} h\Big(x(k|t), \varphi(v(k|t), x(k|t), z(k|t))\Big) \in \mathbb{Y}_t, \quad \forall x(k|t) \in \{z(k|t)\} \oplus \Omega, \ \forall k \in \mathbb{I}_{[0,N-1]}. \tag{4.30}$$

We recall that only the nominal state and input are considered within Problem 4.1. However for satisfaction of the average constraint, all possible real states and inputs in (4.30) must be taken into account, thus, we over-approximate the real states by the RCI set Ω. This additional new constraint would lead to a semi-infinite programming problem (see, e.g., Guerra Vázquez et al. (2008)) and solving this new problem online at each time step can be intractable. Also, determining offline a (constant) tightening $\overline{\mathbb{Y}}$—similar to the tightening of the pointwise-in-time constraints in (2.16)—is in general not possible due to the time-dependency of $\mathbb{Y}_t$.

In order to approach this problem, we use ideas from *robust optimization* (see, e.g., Ben-Tal et al. (2009)). The basic idea is to rewrite an disturbance-affected optimization problem in such a way that the disturbances only occur within the constraints (this is called the *robust counterpart*). Subsequently, one aims to find a solution to this rewritten problem providing a feasible solution for all possible values of the disturbance.

It is known that $e(k|t) = x(k|t) - z(k|t) \in \Omega$ for all possible disturbance sequences. By introducing

$$\overline{h}\Big(z(k|t), v(k|t)\Big) := \max_{e \in \Omega} h\Big(z(k|t) + e, \varphi(v(k|t), z(k|t) + e, z(k|t))\Big), \tag{4.31}$$

it holds that $\overline{h}(z(t), v(t)) \geq h(x(t), u(t))$ for all $t \in \mathbb{I}_{\geq 0}$. By means of this upper bound, we adapt the MPC scheme in order to satisfy $\mathrm{Av}[\overline{h}(\boldsymbol{z}, \boldsymbol{v})] \subseteq \mathbb{R}^p_{\leq 0}$ by the closed-loop nominal sequence. With this, we derive that the desired bound $\mathrm{Av}[h(\boldsymbol{x}, \boldsymbol{u})] \subseteq \mathbb{R}^p_{\leq 0}$ is satisfied by the real closed loop.

Considering the robust optimal steady-state, the new constraints must be taken into account in order to provide feasibility with respect to the average constraints as well. The robust optimal steady-state under average constraints $(z_s^{\mathrm{av}}, v_s^{\mathrm{av}})$ can be derived as in (4.6), however, it must also satisfy $\overline{h}(z_s^{\mathrm{av}}, v_s^{\mathrm{av}}) \in \mathbb{R}^p_{\leq 0}$. This leads to

$$(z_s^{\mathrm{av}}, v_s^{\mathrm{av}}) = \underset{\substack{(z,v) \in \overline{\mathbb{Z}},\, z = f(z,v,0), \\ \overline{h}(z,v) \in \mathbb{R}^p_{\leq 0}}}{\arg\min} \ \ell^{\mathrm{int}}(z, v). \tag{4.32}$$

The minimum exists due to continuity of ℓ and h, compactness of $\mathbb{Z}$, and since $\mathbb{R}^p_{\leq 0}$ is closed.

Because of the average constraint, we must also adapt the assumptions on the terminal ingredients in order to be able to achieve recursive feasibility:

Assumption 4.3. *The terminal region* $\overline{\mathbb{X}}_f \subseteq \overline{\mathbb{X}}$ *is closed and contains* z_s^{av} *in its interior. There exists a local auxiliary controller* $\bar{\kappa}_f : \overline{\mathbb{X}}_f \to \mathbb{U}$ *and a terminal cost* $V_f^{int} : \overline{\mathbb{X}}_f \to \mathbb{R}$ *satisfying Assumption 4.1 (with respect to* (z_s^{av}, v_s^{av})*) as well as for all* $z \in \overline{\mathbb{X}}_f$,

$$\overline{h}\Big(z, \bar{\kappa}_f(z)\Big) \in \mathbb{R}^p_{\leq 0}. \tag{4.33}$$

Remark 4.22. *There is no general procedure to compute the terminal ingredients such that the additional condition* (4.33) *is satisfied. A practical approach is to compute the terminal ingredients, e.g., by means of the method presented in Amrit et al. (2011), and check for satisfaction of* (4.33). *In case the condition is not satisfied, one can always find a feasible terminal condition when replacing the terminal region by an equality constraint on the robust optimal steady-state, i.e.,* $z(N|t) = z_s^{av}$ *replaces* (4.4d) *and* $V_f^{int}(z) \equiv 0$ *in Problem 4.1. However, this usually provides a worse closed-loop (transient) performance. In order to provide at least the same performance, one can adapt the method presented in Müller et al. (2014b, Section 3), where a time-varying terminal region as well as an additional time-varying set* $\mathbb{Y}(t)$ *is added to the recursion* (4.29). *In general, this does not only provide feasibility with respect to the average constraint within the terminal region, but can also lead to a better closed-loop (transient) performance due to additional degrees of freedom.*

Resembling the idea in nominal economic MPC, we introduce an average constraint on the nominal open-loop predictions for each time $t \in \mathbb{I}_{\geq 0}$

$$\sum_{k=0}^{N-1} \overline{h}\Big(z(k|t), v(k|t)\Big) \in \overline{\mathbb{Y}}_t, \tag{4.34}$$

where the set $\overline{\mathbb{Y}}_t$ is provided (similar as in (4.29)) by the recursion

$$\overline{\mathbb{Y}}_{t+1} = \overline{\mathbb{Y}}_t \oplus \mathbb{R}^p_{\leq 0} \oplus \Big\{-\overline{h}\Big(z(0|t), v(0|t)\Big)\Big\}, \qquad \overline{\mathbb{Y}}_0 = \overline{\mathbb{Y}}_{00} \oplus \mathbb{R}^p_{\leq 0}, \tag{4.35}$$

and $\overline{\mathbb{Y}}_{00}$ is an arbitrary compact set containing z_s^{av}.[3] The update in (4.35) is with respect to the over-approximated output $\overline{h}$ and the closed-loop nominal sequence.

Extending Problem 4.1 with the constraint in (4.34), and applying this new problem in Algorithm 4.2, we can state the closed loop

$$x(t+1) = f\Big(x(t), \varphi(v^*(0|t), x(t), z^*(0|t)), w(t)\Big) \tag{4.36a}$$

$$y(t) = h\Big(x(t), \varphi(v^*(0|t), x(t), z^*(0|t))\Big), \tag{4.36b}$$

where $z^*(0|t)$ and $v^*(0|t)$ are provided by the extended problem. The set of all feasible nominal states is denoted by $\overline{\mathbb{X}}_N^{av} := \{z \in \overline{\mathbb{X}} : \exists \, \boldsymbol{v} = \{v(0), \dots, v(N-1)\}$ s.t. $z(0) = z, z(k+1) = f(z(k), v(k), 0), (z(k), v(k)) \in \overline{\mathbb{Z}}, \forall k \in \mathbb{I}_{[0,N-1]}, z(N) \in \overline{\mathbb{X}}_f$, and $\sum_{k=0}^{N-1} \overline{h}\Big(z(k), v(k)\Big) \in \overline{\mathbb{Y}}_0\}$. Using this closed-loop system, we can state the following result:

Theorem 4.23. *Let* $x_0 \in \overline{\mathbb{X}}_N^{av} \oplus \Omega$. *Suppose that Assumptions 2.1, 2.3, 2.7–2.9, 4.2, and 4.3 are satisfied and let* (z_s^{av}, v_s^{av}) *be the robust optimal steady-state satisfying the average constraints according to* (4.32). *Then Problem 4.1 extended with* (4.34) *is feasible*

---

[3] $\overline{\mathbb{Y}}_{00}$ has to be large enough such that (4.34) is initially feasible.

for all $t \in \mathbb{I}_{\geq 0}$. Furthermore, the closed-loop system (4.36) *satisfies the pointwise-in-time constraints* (2.2) *and the average constraints* (4.27). *The closed-loop nominal sequence z^* with its associated input sequence v^* has a robust asymptotic average performance which is no worse than that of the robust optimal steady-state, i.e.,*

$$\limsup_{T \to \infty} \frac{1}{T} \sum_{t=0}^{T-1} \ell^{\text{int}}\Big(z^*(0|t), v^*(0|t)\Big) \leq \ell^{\text{int}}\big(z_s^{\text{av}}, v_s^{\text{av}}\big). \tag{4.37}$$

This result is based on an appropriate tightening of the output function. Computing the upper bounding function $\overline{h}$ can be difficult for general output functions (and input parametrizations). Thus, we discuss some simplifications and approaches to compute $\overline{h}$.

General output functions To simplify the derivation of $\overline{h}$, we consider the output function component-wise and introduce the following assumptions (see Houska (2011)):

Assumption 4.4. *The auxiliary output function h is twice continuously differentiable.*

Assumption 4.5. *For each $i \in \mathbb{I}_{[1,p]}$, there exists a twice continuously differentiable and non-negative function $\widetilde{\lambda}_i : \overline{\mathbb{Z}} \to \mathbb{R}_{\geq 0}$ which satisfies for all $e \in \Omega$*

$$\lambda_{\max}\left(\frac{\partial^2 h_i\Big(z + e, \varphi(v, z + e, z)\Big)}{\partial e^2}\right) \leq 2\widetilde{\lambda}_i(z, v). \tag{4.38}$$

This assumption states that the maximal eigenvalue of the Hessian of each h_i with respect to e is for all $e \in \Omega$ bounded from above by $2\widetilde{\lambda}_i$.

By means of this assumption, we can follow the idea in Houska (2011) and use a Taylor series expansion in order to provide an over-approximation to h by

$$
\begin{aligned}
h_i(x, u) &\leq \max_{e \in \Omega} h_i\Big(z + e, \varphi(v, z + e, z)\Big) \\
&\leq \max_{e, s \in \Omega} \left\{ h_i|_{e=0} + \left.\frac{\partial h_i}{\partial e}\right|_{e=0} e + \frac{1}{2} e^\top \left.\frac{\partial^2 h_i}{\partial e^2}\right|_{e=s} e \right\} \\
&\leq \max_{e \in \Omega} \left\{ h_i|_{e=0} + \left.\frac{\partial h_i}{\partial e}\right|_{e=0} e + \widetilde{\lambda}_i(z, v)\, e^\top e \right\} \\
&=: \widetilde{h}_i(z, v).
\end{aligned}
\tag{4.39}
$$

We can see that for all $e \in \Omega$, it holds that $h(x, u) \leq \overline{h}(z, v) \leq \widetilde{h}(z, v)$. In general, the resulting $\widetilde{h}_i$ might be non-convex functions and, thus, possibly hard to determine. However, in some special cases (depending on the shape of Ω and the properties of h), further simplifications and convexifications can be performed, see, e.g., Houska (2011).

Linear output functions The special class of linear auxiliary output functions of the form

$$h(x, u) = h_{\text{x}} x + h_{\text{u}} u + h_{\text{c}}, \tag{4.40}$$

with $h_{\text{x}} \in \mathbb{R}^{p \times n}$, $h_{\text{u}} \in \mathbb{R}^{p \times m}$, and $h_{\text{c}} \in \mathbb{R}^p$, is investigated in Bayer and Allgöwer (2014) (for linear systems). The basic idea is to use the Pontryagin difference in order to tighten the average constraints appropriately. Considering the limitation to linear systems in Bayer and Allgöwer (2014), the approach can be extended to arbitrary nonlinear systems if the associated RCI error set Ω can be determined by means of a linear input parametrization of the form (2.21), i.e., $\varphi(v, x, z) = v + K(x - z)$. Under this input parametrization, we determine the over-approximation $\overline{h}$ by

$$\overline{h}(z, v) = h_x z + h_u v + h_c + \max_{e \in \Omega} \left\{ h_x e + h_u K e \right\}. \tag{4.41}$$

Using the special structure of this approximation, we introduce the tightened set $\widetilde{\mathbb{Y}} = \mathbb{R}^p_{\leq 0} \ominus h_x \Omega \ominus h_u K \Omega$ and the recursion $\widetilde{\mathbb{Y}}_{t+1} = \widetilde{\mathbb{Y}}_t \oplus \widetilde{\mathbb{Y}} \oplus \{-h(z(t), v(t))\}$ with $\widetilde{\mathbb{Y}}_0 = N\widetilde{\mathbb{Y}} \oplus \widetilde{\mathbb{Y}}_{00}$. Including the additional constraint $\sum_{k=0}^{N-1} h(z(k|t), v(k|t)) \in \widetilde{\mathbb{Y}}_t$ to Problem 4.1 results in a finite-dimensional nonlinear program.

Remark 4.24. *As discussed above, the closed-loop system resulting from application of an economic MPC algorithm does not necessarily converge to a steady-state, but can exhibit some more complex behavior. However, in certain applications, closed-loop convergence is an important requirement which needs to be satisfied and cannot be traded off against an improved performance. In this case, appropriately chosen average constraints can be used to enforce convergence. For nominal economic MPC, several methods were presented, for example in (Angeli et al., 2011; Müller et al., 2014b), how the output function h can be defined such that (asymptotic) convergence of the closed-loop system can be guaranteed. These methods can be transferred to the setting discussed above, i.e., for robust economic MPC, in order to determine suitable functions $\overline{h}$. This way, asymptotic convergence of the nominal system to some desired feasible steady-state $\tilde{z}_s$ can be ensured, and the real closed-loop system then converges to the set $\{\tilde{z}_s\} \oplus \Omega$.*

4.5 Numerical examples

In this section, we apply the presented method of tube-based robust economic MPC to a nonlinear system subject to disturbances. While in the first example we focus on the different performances of the presented approaches, the second example takes average constraints into account.

4.5.1 Nonlinear example

We apply our approach to a nonlinear example which has already been studied in the framework of robust stabilizing MPC (see Limón et al. (2002a) and Bayer et al. (2013)). The dynamics are bilinear and given by

$$x(t + 1) = \begin{bmatrix} 0.55 & 0.12 \\ 0 & 0.67 \end{bmatrix} x(t) + \left(\begin{bmatrix} -0.6 & 1 \\ 1 & -0.8 \end{bmatrix} x(t) + \begin{bmatrix} 0.01 \\ 0.15 \end{bmatrix} \right) u(t) + w(t). \tag{4.42}$$

This system was also considered in Section 3.4.2 for comparing different approaches in min-max robust economic MPC. Here, we use different constraints and a different disturbance set, $\mathbb{X} = \{x \in \mathbb{R}^n : \|x\|_\infty \leq 1\}$ and $\mathbb{U} = \{u \in \mathbb{R} : |u| \leq 0.1\}$. The disturbance set is given by $\mathbb{W} = \{w \in \mathbb{R}^2 : \|w\|_\infty \leq 0.05\}$. As was the case in Section 3.4.2, we can make use of the framework presented in Bayer et al. (2013) to determine an RCI set. With the Lyapunov

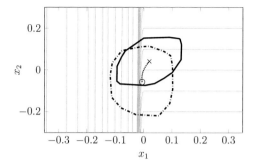

Figure 4.3: Contour plot of the stage cost ℓ in (4.44). All feasible steady-states are given by the dotted line. The convex hull of the closed-loop states are depicted by the sets (Algorithm 4.2 – solid; Algorithm 2.16 – dash-dotted). The optimal steady-states z_s^{int} and x_s are represented by the cross and circle, respectively.

candidate function $V(e) = \|e\|_\infty$, an RCI set for the error dynamics of this system is given by

$$\Omega = \{e \in \mathbb{R}^2 : \|e\|_\infty \leq 1/3\}. \tag{4.43}$$

With this approach, the input for the real system is the same as for the nominal system, that is, $\varphi(v, x, z) = v$. The cost function is given by

$$\ell(x, u) = 0.001\, x_2 + \begin{cases} -100\, x_1^3 & \text{for } x_1 < 0 \\ 0 & \text{for } x_1 \geq 0. \end{cases} \tag{4.44}$$

Using this cost function, one tries to minimize the second state x_2 while keeping the first state x_1 positive by means of a soft constraint. Computing the robust optimal steady-state for this example, we receive $z_s^{\text{int}} = (0.0198, 0.0414)^\top$, the nominal optimal steady-state is located at $x_s = (-0.0055, -0.0578)^\top$. The bound on the asymptotic average performance results in $\ell^{\text{int}}(z_s^{\text{int}}, v_s^{\text{int}}) = 0.1610$.

The performance of the presented tube-based robust economic MPC approach is compared to a tube-based robust MPC approach not considering the disturbance explicitly within the objective, that is, we employ Algorithm 2.16, which takes Problem 2.15 with the economic stage cost ℓ in (4.44) into account. We take a prediction horizon of $N = 20$ and determine the terminal cost, the terminal controller, and the terminal region for both approaches by means of the methodology presented in Amrit et al. (2011) using the stage cost function and the integrated stage cost function, respectively. For the asymptotic average performance of the closed-loop system (averaged over 100 simulations starting at the origin with 200 closed-loop iterations each, and w uniformly distributed over $\mathbb{W}$) follows:

Algorithm 4.2	Algorithm 2.16
0.0010	0.0047

One finds that the asymptotic average performance is significantly improved when using the proposed robust economic MPC scheme based on Algorithm 4.2 compared to the one

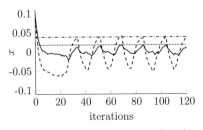 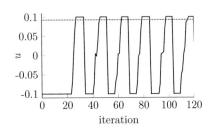

(a) Closed-loop state trajectories x_1 (solid) and (b) Closed-loop input trajectory u (solid) and x_2 (dashed), as well as $z_{s,1}^{\text{av}}$ (dotted) and $z_{s,2}^{\text{av}}$ v_s^{av} (dotted). (dash-dotted).

Figure 4.4: Closed-loop behavior of the real system under average constraints.

using Algorithm 2.16. When considering the convex hull of the closed-loop states depicted in Figure 4.3, one can see the reason for this: For the robust economic MPC approach in Algorithm 4.2 the closed-loop states are kept at a larger x_2-level, which is disadvantageous for the linear part of the cost. However, this approach takes the disturbances into account and avoids closed-loop states with $x_1 < 0$ as this deteriorates the performance. Algorithm 2.16 does not take disturbances into account. Hence, it pushes into x_2-direction as much as possible not incorporating the effects of the disturbance which deviates the closed-loop states to values of x_1 that are smaller than 0.

For this particular example, Algorithm 4.8 achieves the same performance as Algorithm 4.2. This is due to the input, which is at the constraint for both approaches, such that the additional degree of freedom does not redound to the advantage.

4.5.2 Average constraints

In this section, we impose average constraints on the nonlinear system (4.42). The disturbances are restricted to the smaller set $\mathbb{W} = \{w \in \mathbb{R}^2 : \|w\|_\infty \leq 0.005\}$ leading to the RCI set $\Omega = \{e \in \mathbb{R}^2 : \|e\|_\infty \leq 1/30\}$. For this example, we consider a different stage cost $\ell(x, u) = x_2$, which minimizes the value of x_2. Our second goal is satisfaction of two average constraints expressed by the output functions

$$h_1(x) = x_1^2 - 100\, x_2^2 \qquad \text{and} \qquad h_2(u) = -u. \qquad (4.45)$$

The first constraint restricts the states on average to a sector, while the second keeps the input positive on average. For the initial state we set $x_0 = [0.1, 0.1]^\top$ and we choose $\overline{\mathbb{Y}}_{00} = \{y \in \mathbb{R}^2 : \|y\|_\infty \leq 0.6\}$. The input is parametrized according to $\varphi(v, x, z) = v$; the prediction horizon is given by $N = 20$.

We can see that $z_s^{\text{av}} = (0.0180, 0.0385)^\top$ with $v_s^{\text{av}} = 0.0925$ is the robust optimal steady-state including the average constraints as defined in (4.32). Computing the robust optimal steady-state without average constraints as defined in (4.6) ends up at $z_s^{\text{int}} = (-0.0055, -0.0578)^\top$ with $v_s^{\text{int}} = -0.1$.

Considering the simulation results, the closed-loop behavior can be separated into three phases. In the first phase (iteration 0 to 10), the system converges to the robust optimal steady-state $\left(z_s^{\text{int}}, v_s^{\text{int}}\right)$ without average constraints. We note that $\left(z_s^{\text{int}}, v_s^{\text{int}}\right)$ would be feasible including the average constraint expressed by h_1, however, not for the one expressed by h_2. By choosing $\overline{\mathbb{Y}}_{00}$ rather large, it is possible to keep the system for some iterations at the economically best steady-state without violating the average constraint (4.27). After some iterations (around iteration 25), the additional freedom induced by $\overline{\mathbb{Y}}_{00}$ is no longer available. The length of this transient phase can be tuned by the size of $\overline{\mathbb{Y}}_{00}$. Once this additional freedom is no longer available and in order to meet both average constraints, the input must become positive for some iterations. However, the constraints need to be satisfied on average only and, thus, the input can again become negative after a few iterations, see Fig. 4.4b. This leads to a cyclic closed-loop behavior. By means of this cyclic behavior, a closed-loop performance can be achieved, which is better than staying at $\left(z_s^{\text{av}}, v_s^{\text{av}}\right)$, while still satisfying the average constraints.

4.6 Summary

In this chapter, we presented and discussed a second concept to consider information about the disturbance within an economic framework. The approach rests upon the idea that robust control invariant error sets provide an approximation of all states of the real system possible due to the disturbance acting on the system. Within the considered optimal control problem, this is taken into account by integrating the stage cost function over the RCI set. Based on this idea, we showed results on the asymptotic average performance of the closed-loop system. Moreover, optimal steady-state operation and its relation to Ω-robust dissipativity was investigated. We proved that Ω-robust dissipativity is sufficient and (under a certain controllability assumption) necessary for optimal operation at steady-state, and we also provided a connection of this result to the min-max robust economic MPC setup from Chapter 3. Interestingly, it turned out that considering the nominal initial state as an optimization variable has a significant influence on the deducible statements. Asymptotic stability under disturbances was proven based on a strict dissipativity statement. Additionally, we added average constraints to the considered framework and analyzed how these can be handled in the robust setting.

While the closed-loop performance achieved by the tube-based robust economic MPC approach is often better than the one achieved for the min-max approach based on tubes (see, e.g., the numerical example in Section 6.3), there is no general statement. Taking the average over the RCI set is, again, only an approximation of the real distribution of the error over the respective RCI sets. Thus, we take the proper distribution over the error sets into account within the next chapter.

Chapter 5

Improving robust economic MPC using stochastic information

In the previous chapters, we accounted for the disturbances in the economic MPC setup by considering bounds on the error affected systems within the objective of the finite horizon optimal control problems. By doing so, the presented methods inherently take a certain assumption on the disturbance into account. Either only the worst case is considered or we artificially assume all possible errors within the invariant error set to have equal probability. If no further error information is available, these are the most natural and reasonable ways to take disturbances into account. However, if further information about the disturbance is available, for example by means of distribution information of the disturbance, it is beneficial to account for this information within the MPC scheme. Considering the additional information can improve the average performance significantly.

In this chapter, we present how the additional information can contribute within an MPC setup (Section 5.1). Moreover, we provide a constructive approach to find appropriate terminal cost functions (Section 5.2) and investigate optimal operating regimes in this new setup (Section 5.3). Additionally, we present an intermediate approach in between the averaging tube-based robust economic MPC setup presented in Chapter 4 and the approach using stochastic information (Section 5.4).

The results presented in this chapter are based on (Bayer et al., 2015a, 2016a, 2017).

5.1 Problem setup and performance result

In this chapter, we consider linear time-invariant disturbance-affected systems of the form

$$x(t+1) = Ax(t) + Bu(t) + w(t), \qquad x(0) = x_0, \tag{5.1}$$

(equivalent to (2.20)), for which we want to find a feasible control input minimizing the expected asymptotic average cost

$$\limsup_{T \to \infty} \frac{1}{T} \sum_{t=0}^{T-1} \mathbb{E}\left\{ \ell\big(x(t), u(t)\big)\big| x_0 \right\}. \tag{5.2}$$

We note that $x(t)$ is a random variable. With respect to the input, a linear error feedback parametrization of the form (2.21) is assumed, that is, $u(t) = K(x(t) - z(t)) + v(t)$, where $v(t)$ is the manipulated input at time $t \in \mathbb{I}_{\geq 0}$. With regard to this parametrization, we impose the following assumption:

Assumption 5.1. *The closed-loop system matrix $A_{cl} := A + BK$ is invertible and asymptotically stable, i.e., all eigenvalues are strictly inside the unit disc.*

The disturbance is equipped with stochastic information by means of a probability density function (PDF) as in Section 2.2.3. We exploit this additional information within a robust economic MPC scheme. Obviously, one expects the asymptotic average performance to be better than the respective performances determined by means of the robust economic MPC schemes presented in the previous Chapter 3 and 4.

In order to account for the distribution information, the distribution of the error is propagated along the predictions. Following the discussion in Section 2.2.2, the artificial predicted nominal system follows by

$$z(k + 1|t) = Az(k|t) + Bv(k|t), \qquad z(0|t) = x(t), \tag{5.3}$$

where we explicitly set the nominal initial state to be the measured real system state $x(t)$. Using (5.3), we can define the predicted real system

$$x(k + 1|t) = Ax(k|t) + B\big(K(x(k|t) - z(k|t)) + v(k|t)\big) + w(k + t), \qquad x(0|t) = x(t). \tag{5.4}$$

Introducing the predictions for the error by

$$e(k|t) = x(k|t) - z(k|t), \tag{5.5}$$

we can determine the dynamics for the error predictions by

$$e(k + 1|t) = A_{cl}e(k|t) + w(k + t), \qquad e(0|t) = 0. \tag{5.6}$$

As in Section 2.2.3, the error at prediction k is limited to the error set Ω_k determined by means of the recursion (2.23), that is, $\Omega_{k+1} = A_{cl}\Omega_k \oplus \mathbb{W}$, with $\Omega_0 = \{0\}$. Moreover, the limit as $t \to \infty$—denoted by Ω_∞—exists and is the compact mRPI set.

Within robust stabilizing MPC, these error sets have been used in order to reduce conservatism within the predictions as much as possible. In addition, we want to further exploit the stochastic information available about the disturbance provided in Assumption 2.12, namely the PDF $\rho_{\mathbb{W}}$ for the disturbance in $\mathbb{W}$. With this information, one can compute the distribution of the predicted error $e(k|t)$ on the error sets Ω_k. It is a known result in the literature (see, e.g., Klenke (2014, Theorem 2.31)) that the summation of two independent random variables $X, Y \in \mathbb{R}^n$ with their associated PDFs $f_X(\epsilon)$ and $f_Y(\epsilon)$ results in a random variable with PDF

$$f_{X+Y}(\epsilon) = \int_{\mathbb{R}^n} f_X(\epsilon - y) f_Y(y) dy =: (f_X * f_Y)(\epsilon). \tag{5.7}$$

Moreover, we additionally have to consider a linear transformation—due to the system matrix A_{cl}—which is given by

$$f_{A_{cl}X}(\epsilon) = \frac{1}{|\det(A_{cl})|} f_X(A_{cl}^{-1}\epsilon). \tag{5.8}$$

As discussed above, the state $e(k|t)$ of the error system (5.6) is contained in the set Ω_k for all $k \in \mathbb{I}_{\geq 0}$; denote its PDF by $\rho_{\Omega_k}(\epsilon)$. Since $e(0|t) = 0$, $\rho_{\Omega_0}(\epsilon) = \delta(\epsilon)$, where $\delta(\epsilon)$ is the

Dirac delta function. By means of the above, the error can be interpreted as a random variable whose PDF is given by the recursion

$$\rho_{\Omega_{k+1}}(\epsilon) = (\rho_{A_{\mathrm{cl}}\Omega_k} * \rho_{\mathbb{W}})(\epsilon), \tag{5.9}$$

for all $k \in \mathbb{I}_{\geq 0}$, with $\rho_{\mathbb{W}}$ as in Assumption 2.12. With Klenke (2014, Theorem 1.101), we can determine that $\rho_{A_{\mathrm{cl}}\Omega_k}(\epsilon) = 0$ if $\epsilon \notin A_{\mathrm{cl}}\Omega_k$. Using the definition of the convolution, it follows that the support of two convolved distributions is the Minkowski sum of the two supports. This means that the support of the PDF ρ_{Ω_k} of the error at each iteration is equal to the associated error set at this iteration.

Remark 5.1. *Determining the initial state $x(t)$ (exactly) might be a challenging task, for example, due to measurement noise or observer errors. If measurement noise is provided by some PDF $\rho_{\mathbb{V}}$, we can set $\rho_{\Omega_0}(\epsilon) = \rho_{\mathbb{V}}(\epsilon)$, thereby replacing the Dirac delta function. In case of observer errors, ideas similar to Chisci and Zappa (2002) could be of interest extending the approach to account for distributional information. The analysis of this, however, is beyond the scope of this thesis. Thus, we restrict ourselves to the assumption that $x(t)$ can be measured exactly at each time step $t \in \mathbb{I}_{\geq 0}$.*

In order to consider the additional information provided by means of the PDF and in order to overcome the conservatism inherent to the previous methods, the growing error sets as well as the distribution over these sets are taken into account within the finite horizon optimal control problem underlying the MPC algorithm. Therefore, we consider the expected value over the predictions of the stage cost within the objective and tighten the constraints accordingly to be able to guarantee robust constraint satisfaction for the closed-loop system.

Objective   To start with, we introduce and discuss the objective to be used within the finite horizon optimal control problem. As mentioned above, the objective shall consider the expected value within the predictions, that is,

$$\sum_{k=0}^{N-1} \mathbb{E}\left\{\ell\Big(x(k|t), K(x(k|t) - z(k|t)) + v(k|t)\Big)\Big| x(t)\right\} + \overline{V}_{\mathrm{f}}^{\mathrm{int}}\Big(z(N|t)\Big), \tag{5.10}$$

where $\overline{V}_{\mathrm{f}}^{\mathrm{int}}$ is the terminal cost. Subsequently, the abbreviating notation $\mathbb{E}^t\{\cdot\}$ represents the conditional expectation $\mathbb{E}\{\cdot|x(t)\}$.

The terminal cost $\overline{V}_{\mathrm{f}}^{\mathrm{int}}$ can be interpreted as a conditional expectation $\mathbb{E}^t\{\overline{V}_{\mathrm{f}}(x(N|t))\}$ with some appropriate terminal cost $\overline{V}_{\mathrm{f}} : \mathbb{R}^n \to \mathbb{R}$. However, determining $\overline{V}_{\mathrm{f}}^{\mathrm{int}}$ directly is sufficient for the subsequent analysis and turns out to be less demanding than finding an appropriate $\overline{V}_{\mathrm{f}}$.

Regarding the predictions, the only exact information available at time t is the current state $x(t)$. All future states $x(k|t)$, $k \in \mathbb{I}_{[1,N]}$, considered within the prediction are subject to (unknown) disturbances $w(j + t)$, with $j \in \mathbb{I}_{[0,k-1]}$, for which only the bound $\mathbb{W}$ as well as the distribution $\rho_{\mathbb{W}}$ is known. Taking the single terms in (5.10) into account, they can explicitly be computed as

$$\mathbb{E}^t\left\{\ell\Big(x(k|t), K(x(k|t) - z(k|t)) + v(k|t)\Big)\right\} = \int_{\Omega_k} \ell\Big(z(k|t) + \epsilon, K\epsilon + v(k|t)\Big)\rho_{\Omega_k}(\epsilon)\mathrm{d}\epsilon$$
$$=: \ell_k^{\mathrm{int}}\Big(z(k|t), v(k|t)\Big). \tag{5.11}$$

We note that each of the terms ℓ_k^{int} is continuous. This follows by linearity of the integration and by continuity of ℓ (see Assumption 2.3). The terms ℓ_k^{int} can be computed offline for all $k \in \mathbb{I}_{[0,N-1]}$, hence, the objective can be formulated in terms of the nominal state z and input v, only. The objective (5.10) is not time-varying and can be stated a priori. However, for each prediction step $k \in \mathbb{I}_{[0,N-1]}$ within the objective a different cost term must be applied. In the subsequent analysis, we have to consider the stage cost as $k \to \infty$. Thus, we now investigate the terms ℓ_k^{int} in this case.

Lemma 5.2. *Suppose Assumptions 2.2, 2.3, 2.12, and 5.1 are satisfied; let $\ell_k^{\text{int}}(z(k|t), v(k|t))$ denote the conditional expectation of the stage cost ℓ at time $k + t$ given $x(t)$ as defined in (5.11). Then for $k \to \infty$, the conditional expectation ℓ_k^{int} converges, i.e., the limit*

$$\ell_\infty^{\text{int}}(z,v) := \lim_{k\to\infty} \ell_k^{\text{int}}(z,v) \tag{5.12}$$

exists.

Using the convergence result in Lemma 5.2, one can derive the stochastic optimal steady-state (z_s^∞, v_s^∞) as

$$(z_s^\infty, v_s^\infty) = \underset{z=Az+Bv,\,(z,v)\in\overline{\mathbb{Z}}_\infty}{\arg\min} \ell_\infty^{\text{int}}(z,v), \tag{5.13}$$

where $\overline{\mathbb{Z}}_\infty = \mathbb{Z} \ominus (\Omega_\infty \times K\Omega_\infty)$. A stochastic optimal steady-state exists due to compactness of $\overline{\mathbb{Z}}_\infty$ (Assumption 2.11) and continuity of ℓ_∞^{int}, which, in turn, follows by the same arguments as for ℓ_k^{int} and Lemma 5.2. We note that the stochastic optimal steady-state need not be unique. If this is the case, (z_s^∞, v_s^∞) denotes an arbitrary steady-state satisfying (5.13).

Regarding the terminal cost $\overline{V}_{\text{f}}^{\text{int}}$, we introduce the following assumption, based on a terminal region $\overline{\mathbb{X}}_{\text{f}} \subseteq \mathbb{X}$ which will be defined subsequently:

Assumption 5.2. *There exists a terminal cost $\overline{V}_{\text{f}}^{\text{int}} : \overline{\mathbb{X}}_{\text{f}} \to \mathbb{R}$ such that for all $z \in \overline{\mathbb{X}}_{\text{f}}$, the following inequality holds:*

$$\begin{aligned}
\mathbb{E}\left\{ \overline{V}_{\text{f}}^{\text{int}}\left(A_{\text{cl}}z + B(v_s^\infty - Kz_s^\infty) + A_{\text{cl}}^N w \right) \Big| z \right\} - \overline{V}_{\text{f}}^{\text{int}}(z) \\
\leq -\ell_N^{\text{int}}\left(z, K(z - z_s^\infty) + v_s^\infty \right) + \ell_\infty^{\text{int}}(z_s^\infty, v_s^\infty).
\end{aligned} \tag{5.14}$$

Moreover, the terminal cost $\overline{V}_{\text{f}}^{\text{int}}$ is continuous on $\overline{\mathbb{X}}_{\text{f}}$.

We note that a terminal controller of the form $\kappa_{\text{f}}(z) = K(z - z_s^\infty) + v_s^\infty$ is employed within the terminal cost, where $K \in \mathbb{R}^{m \times n}$ is the same as in the error feedback. In principle, a different K could be used within the terminal controller (see Remark 3.13); however, to keep the presentation simple, we stick to the same K in the following discussion. The terminal cost will be of further interest in Section 5.2, where a constructive approach is presented to satisfy Assumption 5.2

Constraints After discussing the objective to be employed, we focus on the constraints, mainly used in order to guarantee constraint satisfaction of the closed-loop system. The techniques used are based on the approach in Chisci et al. (2001), where a tube-based robust stabilizing MPC is presented. As we have already discussed, the objective can be stated as a function of the nominal states and inputs only. Thus, it is desirable to tighten the constraints stating them only in terms of the nominal system.

When introducing the nominal dynamics, the nominal initial state was set to the measured real state $x(t)$ which is assumed to be known exactly. We recall that this means that the initial error prediction is reset at time step t, i.e., $e(0|t) = 0$. In order to meet the pointwise-in-time constraints with the real closed-loop system while only considering the nominal system in the predictions, we must tighten these constraints by taking the error sets Ω_k into account. This leads to the constraint sets

$$\overline{\mathbb{Z}}_k := \mathbb{Z} \ominus (\Omega_k \times K\Omega_k). \tag{5.15}$$

For these sets, Assumption 2.11 is supposed to hold. As is the case for the terms of the stage cost ℓ_k^{int}, the tightened sets $\overline{\mathbb{Z}}_k$ can be computed offline and vary only with respect to the prediction step $k \in \mathbb{I}_{[0,N-1]}$, but are invariant with respect to t.

Considering the terminal constraint, we employ the maximal output admissible set $O_{\max}$ with respect to (z_s^∞, v_s^∞) and $\mathbb{W}$ (see (2.29)) which is used within the terminal region

$$\overline{\mathbb{X}}_f = O_{\max} \ominus \Omega_N. \tag{5.16}$$

We note that this terminal region is robust invariant with respect to any disturbance in the set $A_{\text{cl}}^N \mathbb{W}$ when applying the terminal controller $\kappa_f(z) = K(z - z_s^\infty) + v_s^\infty$.

Combining the presented objective with the introduced constraints, we can state the finite horizon optimal control problem to determine a receding horizon control law at each time $t \in \mathbb{I}_{\geq 0}$ given $x(t)$:

Problem 5.3.

$$\underset{v(t)}{\text{minimize}} \; J_N^{\text{int}}\big(x(t), v(t)\big)$$

subject to

$$z(k+1|t) = Az(k|t) + Bv(k|t), \qquad \forall k \in \mathbb{I}_{[0,N-1]}, \tag{5.17a}$$

$$z(0|t) = x(t), \tag{5.17b}$$

$$\big(z(k|t), v(k|t)\big) \in \overline{\mathbb{Z}}_k, \qquad \forall k \in \mathbb{I}_{[0,N-1]}, \tag{5.17c}$$

$$z(N|t) \in \overline{\mathbb{X}}_f, \tag{5.17d}$$

where

$$J_N^{\text{int}}\big(x(t), v(t)\big) = \sum_{k=0}^{N-1} \ell_k^{\text{int}}\big(z(k|t), v(k|t)\big) + \overline{V}_f^{\text{int}}\big(z(N|t)\big). \tag{5.18}$$

By $V_N^{\text{int}}(x(t))$, we denote the optimal value function, and by $v^*(t) = \{v^*(0|t), \dots, v^*(N-1|t)\}$ the optimal nominal open-loop input sequences at time $t \in \mathbb{I}_{\geq 0}$. As in Section 2.2.3, the set $\widetilde{\mathbb{X}}_N$ denotes the set of all states for which a feasible solution to Problem 5.3 exists. We define the controller by means of the following algorithm:

Algorithm 5.4 (Tube-based robust economic model predictive control using stochastic information). *At each time instant $t \in \mathbb{I}_{\geq 0}$, measure the state $x(t)$ and solve Problem 5.3. Apply the control input $u(t) := v^*(0|t)$ to system* (5.1).

With regard to this algorithm, we state the main result of this section which provides a bound for the asymptotic average performance (5.2):

Theorem 5.5. *Let $x_0 \in \widetilde{\mathbb{X}}_N$. Suppose that Assumptions 2.3, 2.11, 2.12, 5.1, and 5.2 are satisfied and let (z_s^∞, v_s^∞) be the stochastic optimal steady-state according to (5.13). Then, Problem 5.3 is feasible for all $t \in \mathbb{I}_{\geq 0}$. Furthermore, the closed-loop system resulting from applying Algorithm 5.4, that is,*

$$x(t+1) = Ax(t) + Bv^*(0|t) + w(t), \qquad x(0) = x_0, \tag{5.19}$$

satisfies the pointwise-in-time constraints (2.2) for all $t \in \mathbb{I}_{\geq 0}$ and has an expected asymptotic average performance which is no worse than that of the stochastic optimal steady-state, i.e.,

$$\limsup_{T \to \infty} \frac{1}{T} \sum_{t=0}^{T-1} \mathbb{E}\left\{ \ell\big(x(t), v^*(0|t)\big) \big| x_0 \right\} \leq \ell_\infty^{\text{int}}(z_s^\infty, v_s^\infty). \tag{5.20}$$

We want to provide an interpretation for the performance result stated in Theorem 5.5: Each trajectory starting in the mRPI set Ω_∞ centered at the stochastic optimal steady-state will stay inside Ω_∞ if the terminal controller $\kappa_f(x) = K(x - z_s^\infty) + v_s^\infty$ is applied, and will, since the closed-loop state $x(t)$ is a random variable, for $t \to \infty$ be distributed with distribution ρ_{Ω_∞}. Hence, $\ell_\infty^{\text{int}}(z_s^\infty, v_s^\infty)$ can be interpreted as the expected average cost at the stochastic optimal steady-state. This corresponds to the average performance bound $\ell(x_s, u_s)$ typically determined in nominal economic MPC, where (x_s, u_s) is the optimal steady-state (see Section 2.3). Concerning the left hand side in (5.20), the closed-loop trajectory at each time instant t depends on previous disturbances. Because these disturbances are unknown a priori and cannot be taken into account when analyzing the performance, the result must be related to the (known) initial state $x(0) = x_0$ leading to a conditional expectation, which is a known result for example in stochastic (stabilizing) MPC (see Section 2.2.3).

The left hand side of (5.20) is in literature referred to as "ergodic cost" and is, among others, applied in the field of controlling diffusion processes (Borkar, 2005; Borkar and Ghosh, 1988) as well as for efficient exploration methods (Miller et al., 2016).

Remark 5.6. *We note that in order to establish the performance bound (5.20), one cannot just apply results for nominal systems as the one in Amrit et al. (2011) (similar to what was done in Section 4.1). The reason for this is the fact that the cost function ℓ_k^{int} as defined in (5.11) depends on the prediction time k. Hence, the more involved argument considering different expected values as presented in the proof is needed.*

Remark 5.7. *Up to now, we have not discussed the complexity of computing the convolutions in (5.9). In general, this might be a challenging task, especially for higher dimensions. This computation is performed offline, though. It is a standard problem where different schemes exist to handle it, for example, Sample Average Approximation (see, e.g., Pagnoncelli et al. (2009)) or Sparse Grids (see, e.g., Bungartz and Griebel (2004)), to name a few. In most cases, we have to use approximations to compute the PDFs ρ_{Ω_k} in any practical implementation. We note that approximating the PDFs cannot lead to infeasibility of the MPC setup presented, since constraint satisfaction relies on the error sets Ω_k computed in (2.24) only. However, it still may lead to a small performance degradation caused by the approximation error. Using an approximation, we can see when taking the proof into account that the statement in (A.46) need not hold with equality, but an approximation error can occur, which will deteriorate the performance bound. This error can be made arbitrarily small by sufficiently good integration schemes.*

Remark 5.8. *Instead of considering the exact conditional expectation within the objective (5.18), one could consider replacing ρ_{Ω_k} in (5.11) (derived according to the recursion (5.9)) by a uniform distribution over Ω_k. This is comparable to Section 3.2.2, where a similar approach is used in the min-max setup. Such an approach would be computationally appealing since it would eliminate the need of computing the convolution in (5.9). However, the performance bound (5.20) cannot be proven for that setup lacking the essential telescoping series property usually achieved when applying an appropriate candidate solution. An approximation would be needed at this point, which degrades the performance statement.*

Remark 5.9. *In the previous discussion, we considered hard pointwise-in-time constraints on the inputs and states only. Therefore, constraint handling of a tube-based robust stabilizing MPC scheme was adapted in order to provide guarantees. Especially when stochastic information about the disturbance is available, replacing the hard constraints with soft constraints of the form (exemplarily for polytopic constraints on the states)*

$$\mathbb{P}\left\{ [H]_j x(k|t) \leq [h]_j \,|\, x(t) \right\} \leq 1 - [\varepsilon]_j, \qquad j \in \mathbb{I}_{[1,p]}, \ k \in \mathbb{I}_{\geq 0} \tag{5.21}$$

is possible. Here, $\mathbb{P}\{\cdot|x(t)\}$ denotes the conditional probability given $x(t)$ and $Hx \leq h$ is the polytope to be considered consisting of p linear constraints. Thus, the constraint in (5.21) restricts the probability of violating the linear constraint j at time $k+t$ given the realization $x(t)$ to $[\varepsilon]_j$. Constraints of this form can be considered within the robust economic MPC setup by replacing the state and input constraint (5.17c) and the terminal constraint (5.17d) by an appropriate tightening, for example, the tightening presented in Lorenzen et al. (2015).

5.2 Finding an appropriate quadratic terminal cost

In the previous section, we derived a statement bounding the expected asymptotic average performance of the closed loop with respect to the performance at the steady-state. This result is based on Assumption 5.2, where a particular condition is provided for the terminal cost $\overline{V}_{\mathrm{f}}^{\mathrm{int}}$. In general—due to the arbitrary cost function and the distributional information on the disturbances—it is difficult to find a terminal cost satisfying (5.14). A feasible choice satisfying Assumption 5.2 is

$$\overline{V}_{\mathrm{f}}^{\mathrm{int}}\Big(z(N|t)\Big) = \sum_{k=N}^{\infty} \Big(\ell_k^{\mathrm{int}}\Big(z_{\mathrm{f}}(k|t), K(z_{\mathrm{f}}(k|t) - z_s^{\infty}) + v_s^{\infty}\Big) - \ell_{\infty}^{\mathrm{int}}(z_s^{\infty}, v_s^{\infty})\Big), \tag{5.22}$$

with $z_{\mathrm{f}}(k+1|t) = A_{\mathrm{cl}} z_{\mathrm{f}}(k|t) + B(v_s^{\infty} - K z_s^{\infty})$ and $z_{\mathrm{f}}(N|t) = z(N|t)$. However, it can be cumber-some—if not impossible—to find the terminal cost satisfying (5.22) or another terminal cost satisfying (5.14).

In order to overcome this deficiency, we present a systematic and constructive approach to determine a quadratic approximation of the terminal cost. A similar approach was used in Section 3.2.1 in the framework of min-max robust economic MPC based on tubes. We have to point out that using this quadratic approximation slightly deteriorates the a priori provable performance bound (5.20) from Theorem 5.5 as will be shown in the subsequent analysis. The idea of the approach is derived from Amrit et al. (2011), where a quadratic terminal cost is determined in the nominal case. Accordingly, we consider the following assumption:

Assumption 5.3. *The cost function ℓ is twice continuously differentiable on $\mathbb{X} \times \mathbb{U}$.*

For ease of presentation and without loss of generality, the stochastic optimal steady-state is set to zero, i.e., $(z_s^\infty, v_s^\infty) = (0, 0)$.

We pursue the following idea: Instead of using the exact economic stage cost function, we over-approximate this by a quadratic cost function and apply results from linear quadratic stochastic control with this quadratic function. Therefore, one approximates the terminal cost for all nominal states on the terminal region $\overline{\mathbb{X}}_f$ under application of the terminal controller $\kappa_f(z) = Kz$. To this end, we introduce the shifted cost $\bar{\ell}(z) = \ell(z, Kz) - \ell(0, 0)$, and define a quadratic matrix $Q^\star \in \mathbb{R}^{n \times n}$ such that

$$z^\top \left(Q^\star - \bar{\ell}_{zz}(z) \right) z \geq 0, \qquad \forall z \in \overline{\mathbb{X}}_f \oplus \Omega_N, \tag{5.23}$$

where $\bar{\ell}_{zz}$ is the Hessian of $\bar{\ell}$. A matrix $Q^\star$ satisfying (5.23) exists by compactness of $\overline{\mathbb{X}}_f \oplus \Omega_N$ and Assumption 5.3. Due to the disturbances acting on the system, the upper bound for the Hessian must not only be provided for all nominal states within the terminal region, but for all states within the larger region $\overline{\mathbb{X}}_f \oplus \Omega_N$.

Defining the gradient of $\bar{\ell}$ at the stochastic optimal steady-state by $q := \bar{\ell}_z(0)$ and $Q := Q^\star + \alpha I_n$, where $\alpha \in \mathbb{R}_{\geq 0}$ is chosen such that $Q \succ 0$, we can introduce

$$\ell_q(z) = \frac{1}{2} z^\top Q z + q^\top z. \tag{5.24}$$

By the mean value theorem (compare to Amrit et al. (2011, Lemma 23)), it follows that

$$\ell_q(z) - \bar{\ell}(z) = \left(q - \bar{\ell}_z(0) \right)^\top z + \frac{1}{2} \int_0^1 (1 - s) z^\top \left(Q - \bar{\ell}_{zz}(sz) \right) z \, \mathrm{d}s \geq 0, \tag{5.25}$$

and hence,

$$\ell_q(z) \geq \bar{\ell}(z), \qquad \forall z \in \overline{\mathbb{X}}_f \oplus \Omega_N. \tag{5.26}$$

As proposed in Amrit et al. (2011), consider the candidate function

$$\overline{V}_{f,q}^{int}\left(z(N|t) \right) = \sum_{k=N}^{\infty} \ell_q\left(z_f(k|t) \right) = \frac{1}{2} z(N|t)^\top P z(N|t) + p^\top z(N|t), \tag{5.27}$$

where P is the solution to the Lyapunov equation

$$P = A_{cl}^\top P A_{cl} + Q \tag{5.28}$$

and the linear term p satisfies

$$p^\top = q^\top (I_n - A_{cl})^{-1}. \tag{5.29}$$

With the candidate in (5.27) provided, we investigate if this cost satisfies Assumption 5.2. Therefore, we recall that the disturbances are i.i.d. and have zero mean. With results from linear quadratic stochastic control (see, e.g., Bertsekas (2005, Section 4.1), Wang and Boyd (2009)) and with the error dynamics $e(k+1) = A_{cl}e(k) + w(k)$, $e(0) = 0$, it follows that

$$\lim_{k \to \infty} \mathbb{E} \left\{ e(k)^\top Q e(k) \right\} = \mathbb{E} \left\{ w^\top P w \right\}. \tag{5.30}$$

Writing the error in explicit form, $e(k) = \sum_{j=0}^{k-1} A_{\mathrm{cl}}^{k-j-1} w(j)$, we can derive (for $k > N$)

$$
\begin{aligned}
\mathbb{E}\left\{e(k)^\top Q e(k)\right\} &= \sum_{j=0}^{k-1} \mathbb{E}\left\{\left(A_{\mathrm{cl}}^{k-j-1} w(j)\right)^\top Q \left(A_{\mathrm{cl}}^{k-j-1} w(j)\right)\right\} \\
&= \sum_{j=0}^{N-1} \mathbb{E}\left\{\left(A_{\mathrm{cl}}^{N-j-1} w(j)\right)^\top Q \left(A_{\mathrm{cl}}^{N-j-1} w(j)\right)\right\} \\
&\quad + \sum_{j=0}^{k-N-1} \mathbb{E}\left\{\left(A_{\mathrm{cl}}^{k-N-j-1} A_{\mathrm{cl}}^{N} w(j+N)\right)^\top Q \left(A_{\mathrm{cl}}^{k-N-j-1} A_{\mathrm{cl}}^{N} w(j+N)\right)\right\}.
\end{aligned}
\tag{5.31}
$$

Taking the limit as $k \to \infty$ and using (5.30), it follows that

$$
\mathbb{E}\left\{w^\top P w\right\} = \mathbb{E}\left\{e(N)^\top Q e(N)\right\} + \mathbb{E}\left\{\left(A_{\mathrm{cl}}^N w\right)^\top P \left(A_{\mathrm{cl}}^N w\right)\right\}.
\tag{5.32}
$$

To prove the performance bound, we compute the counterpart to (5.14) for the new quadratic terminal cost $\overline{V}_{\mathrm{f,q}}^{\mathrm{int}}$:

$$
\begin{aligned}
&\mathbb{E}\left\{\overline{V}_{\mathrm{f,q}}^{\mathrm{int}}\left(A_{\mathrm{cl}} z + A_{\mathrm{cl}}^N w\right)\Big| z\right\} - \overline{V}_{\mathrm{f,q}}^{\mathrm{int}}(z) \\
&= \frac{1}{2}\left(A_{\mathrm{cl}} z\right)^\top P (A_{\mathrm{cl}} z) - \frac{1}{2} z^\top P z + p^\top A_{\mathrm{cl}} z - p^\top z + \frac{1}{2}\int_{\mathbb{W}}\left(A_{\mathrm{cl}}^N \omega\right)^\top P\left(A_{\mathrm{cl}}^N \omega\right) \rho_{\mathbb{W}}(\omega)\,\mathrm{d}\omega \\
&= -\frac{1}{2}\int_{\Omega_N} (z+\epsilon)^\top Q\,(z+\epsilon)\,\rho_{\Omega_N}(\epsilon)\,\mathrm{d}\epsilon - q^\top z + \frac{1}{2}\int_{\mathbb{W}} \omega^\top P \omega\,\rho_{\mathbb{W}}(\omega)\,\mathrm{d}\omega \\
&\leq -\int_{\Omega_N} \bar{\ell}(z+\epsilon)\rho_{\Omega_N}(\epsilon)\,\mathrm{d}\epsilon + \frac{1}{2}\int_{\mathbb{W}} \omega^\top P \omega\,\rho_{\mathbb{W}}(\omega)\,\mathrm{d}\omega \\
&= -\ell_N^{\mathrm{int}}(z, Kz) + \ell(0,0) + \frac{1}{2}\int_{\mathbb{W}} \omega^\top P \omega\,\rho_{\mathbb{W}}(\omega)\,\mathrm{d}\omega.
\end{aligned}
\tag{5.33}
$$

A few comments with regard to the steps are in order: The second equality follows by adding $\int_{\Omega_N}(-Qz+q)^\top \epsilon\rho_{\Omega_N}(\epsilon)\,\mathrm{d}\epsilon = 0$, which holds due to the disturbance having zero mean, as well as using (5.28), (5.29), and (5.32). The inequality results using (5.24) and (5.26), the last equality from the definition in (5.11).

It turns out that the quadratic terminal cost candidate $\overline{V}_{\mathrm{f,q}}^{\mathrm{int}}$ does not necessarily satisfy Assumption 5.2. Thus, we expect that the bound on the asymptotic average performance provided in (5.20) is not satisfied with the quadratic terminal cost. When performing the same steps as in the proof of Theorem 5.5, we end up with the performance bound

$$
\limsup_{T \to \infty} \frac{1}{T} \sum_{t=0}^{T-1} \mathbb{E}\left\{\ell\big(x(t), v^*(0|t)\big)\Big| x_0\right\} \leq \ell(0,0) + \frac{1}{2}\int_{\mathbb{W}} \omega^\top P \omega\,\rho_{\mathbb{W}}(\omega)\,\mathrm{d}\omega.
\tag{5.34}
$$

However, comparing the right hand sides of (5.20) and (5.34), it follows that

$$
\begin{aligned}
\ell_\infty^{\mathrm{int}}(0,0) &= \int_{\Omega_\infty} \ell(\epsilon, K\epsilon)\rho_{\Omega_\infty}(\epsilon)\,\mathrm{d}\epsilon \\
&\leq \ell(0,0) + \int_{\Omega_\infty} \ell_{\mathrm{q}}(\epsilon)\rho_{\Omega_\infty}(\epsilon)\,\mathrm{d}\epsilon = \ell(0,0) + \frac{1}{2}\int_{\mathbb{W}} \omega^\top P \omega\,\rho_{\mathbb{W}}(\omega)\,\mathrm{d}\omega.
\end{aligned}
\tag{5.35}
$$

In fact, this shows that the bound derived (5.33) with the quadratic terminal cost is a quadratic approximation of the original bound on the average performance in (5.20). Hence, using a quadratic approximation of the terminal cost leads to a degradation of the bound on the asymptotic average performance, namely to a quadratic approximation of the original bound in (5.20).

Remark 5.10. *If a quadratic stage cost function of the form $\ell(x,u) = \frac{1}{2}x^\top Q x + \frac{1}{2}u^\top R u$ with $Q \succ 0$ and $R \succ 0$ is considered, the quadratic terminal cost $\overline{V}_{f,q}^{int}$ satisfies Assumption 5.2, i.e., (5.35) is satisfied with equality. This corresponds to the result for stochastic stabilizing MPC (see Section 2.2.3).*

Remark 5.11. *We want to highlight the similarity of this construction of an appropriate quadratic terminal cost to the approach in Section 3.2.2. The major difference is due to the disturbances, which are assumed to have zero mean in this approach. Thus, no mixed terms in z and w occur, which is a significant simplification for the stochastic approach.*

5.2.1 Improving the a priori bound on the performance

As shown in the section above, deriving a quadratic terminal cost can lead to a conservative bound on the asymptotic average performance. This conservatism is induced by two major reasons: First, the approximation (5.24) must always be a quadratic function with Q *positive definite*. Second, $Q^\star$ must over-approximate the Hessian on the whole set $\overline{\mathbb{X}}_f \oplus \Omega_N$, which might be a large set.

In this section, we present an approach to tackle the second aspect. Therefore, we modify the terminal cost. Again, without loss of generality, $(z_s^\infty, v_s^\infty) = (0,0)$. As discussed at the beginning of Section 5.2, a terminal cost satisfying (5.14) is given by (5.22). However, this cost is computationally impractical. In contrast, we know that the terminal controller κ_f is applied within the closed-loop prediction after N prediction steps. When prolonging the closed-loop predicted trajectory, the nominal states evolve—due to the linear terminal controller—according to

$$z(N + k|t) = A_{cl}^k z(N|t) \in A_{cl}^k \overline{\mathbb{X}}_f. \tag{5.36}$$

Hence, the disturbances affected real states are within the set

$$x(N + k|t) \in \{z(N + k|t)\} \oplus \Omega_{N+k} \subset A_{cl}^k \overline{\mathbb{X}}_f \oplus \Omega_{N+k}. \tag{5.37}$$

By definition of the terminal region, it holds that $A_{cl}^k \overline{\mathbb{X}}_f \oplus \Omega_{N+k} \subseteq \overline{\mathbb{X}}_f \oplus \Omega_N$, due to invariance of the terminal region with respect to $A_{cl}^N \mathbb{W}$. With this, we can restate the terminal cost by

$$\begin{aligned}
\widetilde{V}_{f,q}^{int}\big(z(N|t)\big) = &\sum_{k=0}^{\tilde{N}-1} \ell_{N+k}^{int}\big(z(N + k|t), Kz(N + k|t)\big) \\
&+ \frac{1}{2}z(\tilde{N} + N|t)^\top \tilde{P} z(\tilde{N} + N|t) + \tilde{p}^\top z(\tilde{N} + N|t),
\end{aligned} \tag{5.38}$$

where $\tilde{N} + N \in \mathbb{I}_{>N}$ is the prediction horizon, such that $\tilde{N}$ additional (exact) terms are used in this modified terminal cost function. This is similar to the standard idea in MPC of considering a different prediction (here $N + \tilde{N}$) and control (here N) horizon. Even though this new approach increases the computational effort offline, since we need to compute Ω_{N+1} up to $\Omega_{N+\tilde{N}}$, online only few more evaluations of the stage cost are needed. Moreover, this new approach might lead to an advantage with respect to the quadratic approximation. Instead of requiring a quadratic approximation Q on $\overline{\mathbb{X}}_f \oplus \Omega_N$ as in (5.26), here, the quadratic approximation $\tilde{Q}$ must be a bound only on $A_{cl}^{\tilde{N}} \overline{\mathbb{X}}_f \oplus \Omega_{N+\tilde{N}}$. In general, this set will be much smaller than $\overline{\mathbb{X}}_{\mathrm{f}} \oplus \Omega_N$, and hence, $\widetilde{Q}$ will be a less conservative approximation than Q. Computing $\widetilde{P}$ according to $\widetilde{P} = A_{\mathrm{cl}}^\top \widetilde{P} A_{\mathrm{cl}} + \widetilde{Q}$ accounts for this better approximation within the terminal cost. Following the same steps as in (5.33), it holds that

$$\mathbb{E}\left\{\widetilde{V}_{\mathrm{f,q}}^{\mathrm{int}}\Big(A_{\mathrm{cl}}z + A_{\mathrm{cl}}^N w\Big)\Big| z\right\} - \widetilde{V}_{\mathrm{f,q}}^{\mathrm{int}}\Big(z\Big) \leq -\ell_N^{\mathrm{int}}(z, Kz) + \ell(0,0) + \frac{1}{2}\int_{\mathbb{W}} \omega^\top \widetilde{P}\omega\, \rho_{\mathbb{W}}(\omega)\,\mathrm{d}\omega \quad (5.39)$$

and for the bound on the asymptotic average performance, the result (5.34) follows with P replaced by $\widetilde{P}$. Hence, when $\widetilde{P}$ is smaller than P in the sense that $P - \widetilde{P}$ is positive definite, we can derive a tighter a priori bound on the asymptotic average performance.

5.3 Optimal steady-state operation

In this section, we investigate optimal operation at steady-state of linear disturbance-affected systems in the robust economic MPC framework presented above. The key idea is to consider a stochastic counterpart of dissipativity presented in Definition 2.25 (nominal setup) and Definition 4.6 (tube-based setup). Moreover, we discuss under which conditions a converse result on optimal operation at steady-state, i.e., necessity of dissipativity, can be derived similar to the one in the nominal economic case (Section 2.3) as well as in the tube-based robust economic setup (Section 4.2.3). We show that this, again, relates to a certain controllability and reachability assumption, however, the proof is more involved taking the distributions into account.

First, we define optimal steady-state operation for the setup presented above which considers stochastic information on the disturbance:

Definition 5.12 (Stochastic optimal operation at steady-state)**.** *System* (5.1) *is stochastically optimally operated at steady-state with respect to the cost function ℓ and the constraints* (2.2), *if for each feasible nominal state sequence z, input sequence v, and associated state sequence x, it holds that*

$$\liminf_{T \to \infty} \frac{1}{T} \sum_{t=0}^{T-1} \mathbb{E}\left\{\ell\Big(x(t), K(x(t) - z(t)) + v(t)\Big)\Big| x_0\right\} \geq \ell_\infty^{\mathrm{int}}(z_s^\infty, v_s^\infty), \quad (5.40)$$

where (z_s^∞, v_s^∞) is the stochastic optimal steady-state according to (5.13).

This definition states stochastic optimal operation at steady-state accordingly: There exists no other feasible nominal state and input sequence such that the conditional expectation, given the initial state, of the stage cost is better, i.e., smaller, than the expected average cost at the stochastic optimal steady-state (compare to the discussion after Theorem 5.5). By considering disturbances satisfying Assumption 2.12, that is, disturbances with known distribution, the left hand side of (5.40) can be rewritten and interpreted by the integrated stage cost function introduced in (5.11). For this definition, however, we condition on the initial state $x(0)$, i.e., on information at time $t = 0$, and the provided input parametrization. Thus, the closed-loop average performance achieved can be better than the bound in (5.40) since additional information is taken into account at each time $t \in \mathbb{I}_{>0}$ by means of the measured state $x(t)$. Determining this better bound a priori would require optimization over general feedback laws.

As in the previous setups, we require a certain dissipativity statement in order to guarantee stochastic optimal operation at steady-state. The notion of *stochastic dissipativity* is introduced accordingly (see, e.g., Berman and Shaked (2006) in discrete-time and Wu et al. (2011) in continuous-time):

Definition 5.13 (Stochastic dissipativity). *System* (5.1) *is stochastically dissipative on a set* $\mathbb{S} \subseteq \mathbb{Z}$ *with respect to the supply rate* $s : \mathbb{S} \to \mathbb{R}$ *if there exists a locally integrable and bounded storage function* $\lambda : \mathbb{S}_{\mathbb{X}} \to \mathbb{R}_{\geq 0}$ *such that the following inequality is satisfied for all* $(x, u) \in \mathbb{S}$:

$$\mathbb{E}\left\{ \lambda\big(x(t+1)\big) \big| x(t) \right\} - \lambda\big(x(t)\big) \leq s\big(x(t), u(t)\big). \tag{5.41}$$

We can define the largest set of all state and input pairs in $\overline{\mathbb{Z}}_k$ which are part of a feasible closed-loop sequence by

$$\overline{\mathbb{Z}}_k^o := \Big\{ (x, u) \in \overline{\mathbb{Z}}_k : \exists\, \boldsymbol{v} \text{ s.t. } \big(z(0), v(0) \big) = (x, u),$$
$$z(i+1) = Az(i) + Bv(i), \big(z(i), v(i) \big) \in \overline{\mathbb{Z}}_{k+i}, \forall i \in \mathbb{I}_{\geq 0} \Big\}, \tag{5.42}$$

and its projection onto $\mathbb{X}$ by $\overline{\mathbb{X}}_k^o$.

With these definitions, we can state the main result of this section:

Theorem 5.14. *Consider system* (5.1) *and let Assumptions 2.11, 2.12, and 5.1 be satisfied. Suppose that system* (5.1) *is stochastically dissipative on* $\overline{\mathbb{Z}}_0^o$ *with respect to the supply rate*

$$s(x, u) = \ell(x, u) - \ell_\infty^{\mathrm{int}}(z_s^\infty, v_s^\infty), \tag{5.43}$$

then the system is stochastically optimally operated at steady-state.

5.3.1 Converse result (necessity of dissipativity)

In this section, we investigate necessity of dissipativity for the robust economic MPC setup considering stochastic disturbance information. Similar to the nominal economic and the tube-based robust economic MPC approaches in Section 2.3 and 4.2.3, respectively, necessity of dissipativity is based on a certain controllability and reachability assumption. However, here we need an additional assumption on the convergence of the integrated stage cost function (5.11).

Before discussing the assumptions needed, we first consider the stochastic counterpart of the available storage, which will be of use within the proof. Similar as in Berman and Shaked (2006), the stochastic available storage is introduced by

$$S_{\mathrm{a}}(x) := \sup_{\substack{T \geq 0, \\ z(0)=x, z(k+1)=Az(k)+Bv(k), \\ (z(k),v(k))\in\overline{\mathbb{Z}}_k^o, \forall k \in \mathbb{I}_{\geq 0}}} \sum_{k=0}^{T-1} -\mathbb{E}\left\{ s\big(z(k) + e(k), v(k) + Ke(k) \big) \big| x \right\}, \tag{5.44}$$

recalling that $e(k) = \sum_{j=0}^{k-1} A_{\mathrm{cl}}^{k-j-1} w(j)$. The disturbances $w(j) \in \mathbb{W}$, $j \in \mathbb{I}_{[0,k-1]}$, are distributed according to the PDF $\rho_{\mathbb{W}}$ (see Assumption 2.12). Using the defined integrated stage cost function (5.11), we can reformulate the stochastic available storage as

$$S_{\mathrm{a}}(x) = \sup_{\substack{T \geq 0, \\ z(0)=x, z(k+1)=Az(k)+Bv(k), \\ (z(k),v(k))\in\overline{\mathbb{Z}}_k^o, \forall k \in \mathbb{I}_{\geq 0}}} \sum_{k=0}^{T-1} - \Big(\ell_k^{\mathrm{int}}\big(z(k), v(k) \big) - \ell_\infty^{\mathrm{int}}(z_s^\infty, v_s^\infty) \Big). \tag{5.45}$$

We consider the following assumption on the constraints, which is only a little restriction for most practical applications:

Assumption 5.4. *The set $\mathbb{Z}$ is a compact and convex set. The set $\overline{\mathbb{X}}_0^o$ is a compact and convex set with non-empty interior.*

Regarding the stochastic available storage, we introduce a preliminary result which will be useful in the subsequent discussion.

Lemma 5.15. *Suppose that Assumptions 2.3, 2.11, 2.12, 5.1, and 5.4 are satisfied and that the stochastic available storage $S_a(x)$ is bounded on $\overline{\mathbb{X}}_0^o$. Then, $S_a(x)$ is Riemann integrable on $\overline{\mathbb{X}}_0^o$.*

The stochastic available storage is a useful tool to determine stochastic dissipativity of a system as stated in the following result:

Lemma 5.16. *Suppose that Assumptions 2.3, 2.11, 2.12, 5.1, and 5.4 are satisfied. For any $T \geq 0$, system (5.1) is stochastically dissipative on $\overline{\mathbb{Z}}_0^o$ with supply rate s defined in (5.43) if and only if the stochastic available storage $S_a(x)$ is bounded on $\overline{\mathbb{X}}_0^o$. Moreover, $S_a(x)$ itself is a possible storage function satisfying (5.41).*

Remark 5.17. *In principle, one could relax boundedness of the storage function in Definition 5.13 as well as of the available storage in Lemma 5.16 to finiteness only (compare Berman and Shaked (2006)). However, we need to guarantee (at least) local integrability of the storage function and accordingly of the stochastic available storage $S_a(x)$ for all $x \in \overline{\mathbb{X}}_0^o$. In order to guarantee Riemann integrability (as provided in Lemma 5.15), boundedness of the available storage must be assumed, and hence, we present all the results in terms of bounded storage functions.*

As mentioned before and similar to the previous investigations on necessity of dissipativity, the result is based on certain controllability and reachability conditions provided by the following sets. In contrast to the nominal case (see Section 2.3) and the tube-based case (see Section 4.2.3), we have to consider time-varying tightening of the constraints in order to account for the time-variant error sets (5.6). First, we introduce the set of states which can be controlled to the stochastic optimal steady-state in M steps:

$$\overline{\mathcal{C}}_M := \{ x \in \mathbb{R}^n : \exists \boldsymbol{v} \text{ s.t. } z(0) = x, z(k+1) = Az(k) + Bv(k), z(M) = z_s^\infty,$$
$$\left(z(k), v(k) \right) \in \overline{\mathbb{Z}}_k, \ \forall k \in \mathbb{I}_{[0,M-1]} \}. \tag{5.46}$$

Second, the set of all nominal states that can be reached from the stochastic optimal steady-state in M steps is provided by

$$\overline{\mathcal{R}}_M := \{ x \in \mathbb{R}^n : \exists \boldsymbol{v} \text{ s.t. } z(0) = z_s^\infty, z(k+1) = Az(k) + Bv(k), z(M) = x,$$
$$\left(z(k), v(k) \right) \in \overline{\mathbb{Z}}_k, \ \forall k \in \mathbb{I}_{[0,M]} \}. \tag{5.47}$$

Third, we define the set of nominal state/input pairs which are part of a feasible state/input sequence pair $(\boldsymbol{z}, \boldsymbol{v})$ guaranteeing the nominal state sequence to stay in the intersection of $\overline{\mathcal{C}}_M$ and $\overline{\mathcal{R}}_M$ (compare to (2.42) and (4.19)) by

$$\overline{\mathcal{Z}}_M := \{ (x,u) \in \overline{\mathbb{Z}}_\infty : \exists \boldsymbol{v} \text{ s.t. } \left(z(0), v(0) \right) = (x,u), z(k+1) = Az(k) + Bv(k),$$
$$\left(z(k), v(k) \right) \in \overline{\mathbb{Z}}_\infty, z(k) \in \overline{\mathcal{C}}_M \cap \overline{\mathcal{R}}_M, \ \forall k \in \mathbb{I}_{\geq 0} \}. \tag{5.48}$$

For this set, it holds that $\overline{\mathcal{Z}}_M \subseteq \overline{\mathbb{Z}}_\infty^o$. In accordance to the previous notation, the projection of $\overline{\mathcal{Z}}_M$ onto $\mathbb{X}$ is denoted by $\overline{\mathcal{X}}_M := \{x \in \mathbb{X} : \exists u \in \mathbb{U} \text{ s.t. } (x, u) \in \overline{\mathcal{Z}}_M\}$.

Remark 5.18. *We note that in the definition of $\overline{\mathcal{Z}}_M$, a tightening is considered with respect to Ω_∞ instead of with respect to a less restrictive "growing tube" Ω_k. This is due to the special structure of proof employed later. There a cyclic state/input sequence of finite period is required which one wants to use repeatedly. By taking the conditional expectation of the stage cost into account and letting the time grow to infinity, one needs to restrict each of the elements of these cyclic sequences to $\overline{\mathbb{Z}}_\infty$ in order to guarantee feasibility at all times.*

Remark 5.19. *Considering Lemma 5.15 and 5.16, one can recast the same result when replacing $\overline{\mathbb{Z}}_0^o$ and $\overline{\mathbb{X}}_0^o$ by $\overline{\mathcal{Z}}_M$ and $\overline{\mathcal{X}}_M$, respectively. By Assumption 5.4, $\mathbb{Z}$ is a compact and convex set. Thus, since all additional constraints induced by $\overline{\mathcal{Z}}_M$—thereby one also has to consider $\overline{\mathcal{R}}_M$ and $\overline{\mathcal{C}}_M$—are convex, the same proofs remain valid with the respective sets.*

In order to prepare the main result of this section, we introduce the following assumption on the convergence of the stage cost function:

Assumption 5.5. *Given the system* (5.1) *and the stage cost function ℓ, for any feasible nominal trajectory satisfying $(z(k), v(k)) \in \overline{\mathbb{Z}}_\infty \ \forall k \in \mathbb{I}_{\geq 0}$, there exists a $c < \infty$ such that*

$$\sum_{k=0}^{\infty} \left| \ell_k^{\text{int}}\big(z(k), v(k)\big) - \ell_\infty^{\text{int}}\big(z(k), v(k)\big) \right| \leq c. \tag{5.49}$$

We discuss a rather general condition on the stage cost function ℓ under which this assumption can be satisfied subsequently. The main result of this section states necessity of dissipativity in the considered framework with stochastic information about the disturbance incorporated in the robust economic MPC approach.

Theorem 5.20. *Consider system* (5.1) *and let Assumptions 2.3, 2.11, 2.12, 5.1, 5.4, and 5.5 be satisfied, Suppose that system* (5.1) *is stochastically optimally operated at steady-state. Then, for each $M \in \mathbb{I}_{\geq 1}$, system* (5.1) *is stochastically dissipative on $\overline{\mathcal{Z}}_M$ with respect to the supply rate* (5.43).

With this result, we can see that while stochastic dissipativity is sufficient for stochastic optimal operation at steady-state, the converse result holds true only under a certain controllability/reachability assumption as well as an assumption on the convergence of the integrated stage cost ℓ_k^{int} (see (5.11)) as $k \to \infty$. If Assumption 5.5 is satisfied, it follows similar to Müller et al. (2015a, Corollary 1) and Section 4.2.3 that stochastic dissipativity on $\overline{\mathbb{Z}}_\infty^o$ is necessary and sufficient for stochastic optimal operation at steady-state if $\overline{\mathcal{Z}}_M = \overline{\mathbb{Z}}_\infty^o$ for some $M \in \mathbb{I}_{\geq 1}$. Next, we focus on the convergence of the stage cost function required in Assumption 5.5.

Satisfying Assumption 5.5

In Theorem 5.20, convergence of the integrated stage cost as introduced in Assumption 5.5 is crucial for proving the converse result on necessity of stochastic dissipativity. Now, we investigate a rather general class of stage cost functions ℓ for which Assumption 5.5 holds.

Proposition 5.21. *Suppose that the stage cost function ℓ is a polynomial function of finite degree $\bar{r}$, i.e., $\bar{r} < \infty$ and that Assumptions 2.11, 2.12, and 5.1 are satisfied. Then, Assumption 5.5 is satisfied.*

Summary

We have seen that necessity of dissipativity for optimal operation at steady-state can also be derived in the stochastic setup. Interestingly, in addition to the previously considered controllability assumption, we need an additional assumption on the stage cost function in the stochastic setup. This assumption guarantees that the integrated stage cost function representing the expected value converges "fast enough" to its limit. Additionally, we have seen that a non-trivial class of stage cost functions satisfies this assumption. This indicates that Assumption 5.5 is not too restrictive for many stage cost functions of practical interest.

Moreover, as discussed in the previous Section 2.3.1 and 4.2.3, the converse result (under the convergence and the mild controllability assumption) provides that stochastic dissipativity is a useful and not too conservative characterization for stochastic optimal operation at steady-state.

5.4 Tube-based robust economic MPC with RPI-distribution

In the previous sections, we have made use of the stochastic information about the disturbance within PDFs of the error along open-loop predictions. This can improve the performance compared to the robust economic MPC schemes presented in the previous chapters. However, by considering the error sets which are varying with prediction time, we have a different structure of the underlying optimal control problem. Among others, the approach in Section 5.1 might be computationally more expensive in the sense that different stage cost functions and different tightened sets must be considered at each prediction step. In contrast, the approaches in Section 3.2 and Chapter 4 consider the same stage cost function and tightening for each prediction step.

In this section, we introduce another approach employing the stochastic information of the disturbance while possessing the same complexity as the approaches in Section 3.2 and Chapter 4. This new approach can be seen as an intermediate solution of these approaches.

In particular, and in contrast to the objective (5.18), we consider the distribution of the error over the whole invariant error set Ω_∞. This leads to the following finite horizon optimal control problem to determine a receding horizon control law for each time $t \in \mathbb{I}_{\geq 0}$ given $x(t)$:

Problem 5.22.

$$\underset{\boldsymbol{v}(t)}{\text{minimize}} \; J_{N,\infty}^{\text{int}}\Big(z(0|t), \boldsymbol{v}(t)\Big)$$

subject to

$$z(k+1|t) = Az(k|t) + Bv(k|t), \qquad \forall k \in \mathbb{I}_{[0,N-1]}, \qquad (5.50a)$$

$$z(0|t) = \begin{cases} x(0), & \text{for } t = 0, \\ z(1|t-1), & \text{for } t \in \mathbb{I}_{\geq 1}, \end{cases} \qquad (5.50b)$$

$$\Big(z(k|t), v(k|t)\Big) \in \overline{\mathbb{Z}}_\infty, \qquad \forall k \in \mathbb{I}_{[0,N-1]}, \qquad (5.50c)$$

$$z(N|t) \in \overline{\mathbb{X}}_{\text{f}}, \qquad (5.50d)$$

where

$$J_{N,\infty}^{\text{int}}\Big(z(0|t),\boldsymbol{v}(t)\Big) = \sum_{k=0}^{N-1} \ell_{\infty}^{\text{int}}\Big(z(k|t),v(k|t)\Big) + V_{\text{f},\infty}^{\text{int}}\Big(z(N|t)\Big). \tag{5.51}$$

We note that in contrast to the previous problems, the nominal initial state is fixed by (5.50b). This is a similar approach as described in Section 4.2.1. Here, we need to fix the nominal initial state in order to preserve the information on the initial state. If one leaves the nominal initial state as an optimization variable with a constraint of the form (4.4b), the optimization could set $z(0|t)$ such that $x(t)$ is at the edge of $\{z(0|t)\} \oplus \Omega_\infty$ without accounting for the probability of this event.

Considering the terminal region $\overline{\mathbb{X}}_{\text{f}}$ in (5.50) and the terminal cost $V_{\text{f},\infty}^{\text{int}}$ in (5.51), we introduce the following assumption, which is similar to Assumption 4.1:

Assumption 5.6. *The terminal region $\overline{\mathbb{X}}_{\text{f}} \subseteq \overline{\mathbb{X}}_\infty$ is closed and contains z_s^∞ in its interior and there exists a continuous terminal cost $V_{\text{f},\infty}^{\text{int}} : \overline{\mathbb{X}}_{\text{f}} \to \mathbb{R}$ such that for the given auxiliary control law $\kappa_{\text{f}}(z) = K(z - z_s^\infty) + v_s^\infty$ and for all $z \in \overline{\mathbb{X}}_{\text{f}}$ it holds that*

(i) $\Big(z, K(z - z_s^\infty) + v_s^\infty\Big) \in \overline{\mathbb{Z}}_\infty,$

(ii) $A_{\text{cl}}z + B(v_s^\infty - Kz_s^\infty) \in \overline{\mathbb{X}}_{\text{f}},$ *and*

(iii) $V_{\text{f},\infty}^{\text{int}}\Big(A_{\text{cl}}z + B(v_s^\infty - Kz_s^\infty)\Big) - V_{\text{f},\infty}^{\text{int}}(z) \leq -\ell_\infty^{\text{int}}\Big(z, K(z - z_s^\infty) + v_s^\infty\Big) + \ell_\infty^{\text{int}}(z_s^\infty, v_s^\infty).$

As noted in Remark 2.23, $z_s^{\text{int}} \in \text{int}(\overline{\mathbb{X}}_{\text{f}})$ is only necessary if asymptotic stability is investigated, otherwise $z_s^{\text{int}} \in \overline{\mathbb{X}}_{\text{f}}$ is sufficient. In case of $\ell \in \mathcal{C}^2$, the constructive approach presented in Amrit et al. (2011) can be adapted to the stage cost function ℓ_∞^{int} used in Problem 5.22 in order to find a terminal cost satisfying Assumption 5.6.

Using Problem 5.22, we define the controller by the following algorithm:

Algorithm 5.23 (Tube-based robust economic model predictive control with fixed distribution). *At each time instant $t \in \mathbb{I}_{\geq 0}$, measure the state $x(t)$; for $t \in \mathbb{I}_{\geq 1}$ compute $z(1|t-1) = Az(0|t-1) + Bv^*(0|t-1)$; solve Problem 5.22. Apply the control input $u(t) := v^*(0|t)$ to system (5.1).*

For the closed-loop system resulting from applying Algorithm 5.23, recursive feasibility as well as constraint satisfaction for the constraints (2.2) can be guaranteed. Moreover, it holds for the asymptotic average performance

$$\limsup_{T \to \infty} \frac{1}{T} \sum_{t=0}^{T-1} \ell_\infty^{\text{int}}\Big(z(0|t),v^*(0|t)\Big) \leq \ell_\infty^{\text{int}}(z_s^\infty,v_s^\infty). \tag{5.52}$$

The proof for these results follows along the lines of the proof of Theorem 3.9.

At first sight, the performance result in (5.52) is only a statement for the nominal system, however, the real system is known to stay within the mRPI set around the nominal states. The left hand side can be interpreted as an average performance result for the real closed-loop system, averaged over all possible disturbances by integrating the cost over the mRPI set Ω_∞ according to the asymptotic distribution ρ_{Ω_∞}. For the right hand side, as discussed in Section 5.1, $\ell_\infty^{\text{int}}(z_s^\infty,v_s^\infty)$ can be interpreted as the expected average cost at the stochastic optimal steady-state. This result is comparable to those derived for the bound on the asymptotic average performance in Chapter 4, but with a different way to handle the disturbances within the stage cost function.

Remark 5.24. *Instead of considering optimal operating regimes or asymptotic stability for this particular approach, we refer to the concept of robust dissipativity introduced in Section 4.2.1 and 4.3.1. Problem 5.22 has an equivalent structure to the problem described in Section 4.2.1. The two problems only differ in the considered objective. Hence, the same results hold when replacing ℓ^{int} with $\ell^{\mathrm{int}}_\infty$ and the stochastic optimal steady-state in the supply rate.*

5.5 Summary

In this chapter, we presented two approaches for linear systems to consider stochastic information about the disturbance within the framework of robust economic MPC. The basic idea of the first approach is to consider exact bounds for the error along the prediction horizon. This leads to growing error sets which can be equipped with information about the distribution of the error using standard tools. For this approach, bounds on the expected asymptotic average performance were derived, which are based on appropriate terminal cost functions. In order to provide quadratic terminal cost functions, constructive approaches were presented, however, resulting in a deterioration of the provable bound on the expected asymptotic average performance. Similar to the previous approaches, we were able to prove sufficiency and (under certain controllability and convergence assumptions) necessity of stochastic dissipativity for stochastic optimal operation at steady-state. The second approach considered is an intermediate approach in between the first approach of this chapter and the approach of Chapter 4 which equips the respective mRPI set with information about the steady-state distribution. For this approach, we investigated the asymptotic average performance and showed its connection to the tube-based robust economic MPC approach when facing optimal steady-state operation and asymptotic stability.

Chapter 6

Comparison of the approaches

After having presented different approaches on robust economic MPC separately, we compare them to each other in this chapter. To start with, we introduce another notion of optimality which is solely based on the dynamics, the stage cost, and the constraints. This concept of optimal operation is based on comparison functions and turns out to be the least restrictive optimal operating regime, while also providing the weakest statement (see Section 6.1). Furthermore, we compare the different notions of optimal steady-state operation in a discussion which is substantiated with a numerical example (Section 6.2). Moreover, the numerical simulation of a (simplified) continuous-stirred tank reactor is provided, which allows us to compare the performance for all provided approaches as well as their bounds, and for which we present an elaborate discussion aiming at explaining the different performance results (Section 6.3).

The results of Section 6.1 and 6.2 are based on Bayer et al. (2017). The numerical example in Section 6.3 was also investigated in (Bayer et al., 2014b, 2015a, 2016a,b).

6.1 Optimality based on comparison functions

The previously presented optimal operating regimes in the context of robust economic MPC in Section 4.2 and 5.3 were based on some behavior of the underlying (artificial) nominal system. As discussed in these sections, it turns out that in contrast to Definition 2.24, the chosen algorithm for determining the nominal (and thus the real) closed loop has a significant influence on statements that can be derived concerning the optimal operating behavior. In nominal economic MPC, the chosen algorithm does not influence the optimal operating behavior. Thus, we want to derive a similar optimality statement in the robust case in this section, which is independent of the underlying MPC algorithm and which takes only information about the size of the disturbance set $\mathbb{W}$ into account. When the underlying algorithm is not explicitly taken into account within the analysis, the derivable statement is in general weaker compared to any of the previous notions examining optimality with respect to the MPC algorithm.

In order to gain independence of the underlying MPC algorithm, we introduce a notion of optimality being based on comparison functions (see, e.g., Kellett (2014)). The idea is to determine whether a system will stay in a neighborhood of the desired, optimal equilibrium. We recall $w_{\max} = \max_{w \in \mathbb{W}} |w|$.

Before starting with the optimality discussion itself, we introduce a behavior to compare the closed-loop performance to. Therefore, we state the *optimal steady-state* similar to (2.33) by

$$\Big(x_s(w_{\max}), u_s(w_{\max})\Big) = \underset{z=f(z,v,0),\,(z,v)\in\overline{\mathbb{Z}}(w_{\max})}{\arg\min} \ell(z,v). \tag{6.1}$$

If $w_{\max} = 0$, that is, no disturbance acts on the system, this definition boils down to (2.33). In order to be able to guarantee feasibility, we consider a set tightening

$$\overline{\mathbb{Z}}(w_{\max}) := \Big\{(z,v) \in \mathbb{R}^n \times \mathbb{R}^m : \big(x,v\big) \in \mathbb{Z} \text{ for all } x \in \{z\} \oplus \Omega(w_{\max})\Big\}. \tag{6.2}$$

This is equivalent to the tightening in (2.16) as well as in (3.16) stating explicitly the relation on $w_{\max}$, but without an input parametrization. The RCI set $\Omega(w_{\max})$ for the associated error dynamics (3.15) (again without a parametrization of the input) depends on $\mathbb{W}$. Similar to the discussion in the previous chapters, a steady-state does not exist for the real system under unknown, varying disturbances. The set $\{x_s(w_{\max})\} \oplus \Omega(w_{\max})$ can be interpreted as a robust counterpart of the steady-state, since the state of the real system $x(t)$ remains within this set for all times $t \in \mathbb{I}_{\geq 0}$ when starting in this set and applying $u_s(w_{\max})$ at each iteration. All states in $\{x_s(w_{\max})\} \oplus \Omega(w_{\max})$ are feasible because of the tightening in (6.2).

Remark 6.1. *Instead of computing the optimal steady-states* (6.1) *with respect to* (6.2), *one could also consider the tightening* (3.12) *(without an input parametrization). These steady-states provide the same property as discussed above, namely feasibility of the real system for all possible disturbances when starting at $x_s(w_{\max})$ and applying $u_s(w_{\max})$ at each iteration. However, with respect to Proposition 3.7 and for ease of presentation, the slightly more conservative tightening* (6.2) *is taken into consideration.*

If an input parametrization is required in order to determine an RCI error set, we refer to the parametrization (3.3) along with the associated Definition 3.5 and the tightened constraint set provided in (3.16).

Since the RCI set is a function of $w_{\max}$, so is the tightening. Existence of the optimal steady-state (6.1) follows from Assumptions 2.1–2.3 and 2.7, assuming that $\overline{\mathbb{Z}}(w_{\max})$ is non-empty (compare to Assumption 2.9). Subsequently, $\overline{\mathbb{X}}(w_{\max})$ is the projection of $\overline{\mathbb{Z}}(w_{\max})$ onto $\mathbb{X}$. We note that the minimizer of (6.1) need not be unique for a given $w_{\max}$. If this is the case, one of the minimizers can be chosen.

We define a new notion of dissipativity:

Definition 6.2 ($\mathcal{K}$-robust dissipativity). *System* (2.7) *is $\mathcal{K}$-robustly dissipative on a set $\mathbb{S} \subseteq \mathbb{Z} \times \mathbb{W}$ with respect to the supply rate $s : \mathbb{Z} \to \mathbb{R}$ if there exists a storage function $\lambda : \mathbb{S}_{\mathbb{X}} \to \mathbb{R}_{\geq 0}$ such that for all $(x,u,w) \in \mathbb{S}$*

$$\lambda\Big(f(x,u,w)\Big) - \lambda(x) \leq s(x,u) + \rho(w_{\max}) \tag{6.3}$$

holds, where ρ is a $\mathcal{K}_\infty$ function.

We can now associate $\mathcal{K}$-robust dissipativity with a new definition of robust optimal operation of a system:

Definition 6.3 ($\mathcal{K}$-robust optimal operation at steady-state). *System* (2.7) *is said to be $\mathcal{K}$-robustly optimally operated at steady-state with respect to the stage cost ℓ, the* constraints (2.2), and the disturbance set $\mathbb{W}$, if for each feasible *input and disturbance sequence* $\boldsymbol{u}$ *and* $\boldsymbol{w}$ *and associated state sequence* $\boldsymbol{x}$, *it holds that*

$$\liminf_{T \to \infty} \frac{1}{T} \sum_{k=0}^{T-1} \ell\big(x(k), u(k)\big) \geq \ell\big(x_s(w_{\max}), u_s(w_{\max})\big) - \rho(w_{\max}), \tag{6.4}$$

where $(x_s(w_{\max}), u_s(w_{\max}))$ *is the optimal steady-state introduced in* (6.1) *and* ρ *is a* $\mathcal{K}_\infty$ *function.*

This notion of optimal operation does in particular not depend on the choice of the MPC algorithm. We note that this definition reduces to Definition 2.24 if $w_{\max} = 0$.

Definition 6.3 means that operating the system at the optimal steady-state (6.1) results in an asymptotic average performance being approximately optimal up to an error term depending on the largest disturbance $w_{\max}$.

We can define the largest set of all state, input, and disturbance triples in $\mathbb{Z}$ which are part of a feasible closed-loop sequence

$$\overline{\mathbb{Z}}_{\mathcal{K}}^o := \Big\{ (x, u, w) \in \mathbb{R}^n \times \mathbb{R}^m \times \mathbb{R}^q : \exists \, \boldsymbol{v}, \boldsymbol{\omega} \text{ s.t. } \big(z(0), v(0), \omega(0)\big) = (x, u, w),$$
$$z(k+1) = f\big(z(k), v(k), \omega(k)\big), \big(z(k), v(k)\big) \in \mathbb{Z}, \omega(k) \in \mathbb{W}, \forall k \in \mathbb{I}_{\geq 0} \Big\}. \tag{6.5}$$

Theorem 6.4. *Consider system* (2.7) *and let Assumption 2.7 be satisfied. Suppose that system* (2.7) *is* $\mathcal{K}$-*robustly dissipative on* $\overline{\mathbb{Z}}_{\mathcal{K}}^o$ *with respect to the supply rate*

$$s(x, u) = \ell(x, u) - \ell\big(x_s(w_{\max}), u_s(w_{\max})\big), \tag{6.6}$$

then the system is $\mathcal{K}$-*robustly optimally operated at steady-state* $x_s(w_{\max})$ *defined via* (6.1).

Relating this new notion with the ones presented before, we can see that this is a rather weak statement. In particular, we are only able to state that the steady-state operation is approximately optimal up to an error term depending on the largest disturbance. Additionally, the optimal steady-state (6.1) does not consider the disturbance other than by means of the constraints. As discussed in Chapter 1, this steady-state might not be desirable to operate the system at, since large disturbances can lead to a large value of $\rho(w_{\max})$, i.e., a large deterioration in (6.4). Hence, the information provided by this notion is limited, especially when compared to the other notions of optimal steady-state operation.

For most systems it is possible to find a ρ satisfying (6.3), and moreover, the statement is completely independent of the underlying MPC algorithm and does only depend on the problem setup considered. Thus, it is possible to provide some intuition about the optimal operating behavior at the optimal steady-state (6.1), approximating optimality up to an error term depending on the disturbance.

6.1.1 Converse result (necessity of dissipativity)

We have seen in the previous discussions on optimal operation at steady-state, see Section 2.3, 4.2.3, and 5.3.1, that a converse statement for proving necessity of dissipativity for optimal operation at steady-state is based on specific controllability assumptions. In this section, we investigate a similar statement for the new notion of $\mathcal{K}$-robust dissipativity.

Similar to Müller et al. (2015a) and to our approach in Section 4.2.3, the following definitions provide a controllability/reachability condition. In contrast to these previous approaches, we additionally have to consider the disturbance acting on the system by choosing it together with the inputs.

The set of all states for which input and disturbance sequences exist such that they can be controlled to the optimal steady-state $x_s(w_{\max})$ within M steps is introduced by

$$\overline{\mathcal{C}}_M^{\mathcal{K}} := \Big\{ x \in \mathbb{R}^n : \exists \boldsymbol{v}, \boldsymbol{\omega} \text{ s.t. } z(0) = x, z(k+1) = f\big(z(k), v(k), \omega(k)\big), \tag{6.7}$$
$$z(M) = x_s(w_{\max}), \big(z(k), v(k)\big) \in Z, \omega(k) \in \mathbb{W}, \ \forall k \in \mathbb{I}_{[0,M-1]} \Big\},$$

and the set of states for which input and disturbance sequences exist such that they can be reached from $x_s(w_{\max})$ in M steps by

$$\overline{\mathcal{R}}_M^{\mathcal{K}} := \Big\{ x \in \mathbb{R}^n : \exists \boldsymbol{v}, \boldsymbol{\omega} \text{ s.t. } z(0) = x_s(w_{\max}), z(k+1) = f\big(z(k), v(k), \omega(k)\big), \tag{6.8}$$
$$z(M) = x, \big(z(k), v(k)\big) \in Z, \omega(k) \in \mathbb{W}, \ \forall k \in \mathbb{I}_{[0,M]} \Big\}.$$

We note that here, the disturbance is a free variable, that is, the sets are defined by determining *one* disturbance sequence.

The set of state/input/disturbance triples which are part of a feasible state/input/disturbance sequence triple $(\boldsymbol{z}, \boldsymbol{v}, \boldsymbol{\omega})$ guaranteeing the respective state sequence to stay in the intersection of $\overline{\mathcal{C}}_M^{\mathcal{K}}$ and $\overline{\mathcal{R}}_M^{\mathcal{K}}$ for all times is introduced by

$$\overline{\mathcal{Z}}_M^{\mathcal{K}} := \Big\{ (x, u, w) \in \mathbb{R}^n \times \mathbb{R}^m \times \mathbb{R}^q : \exists \boldsymbol{v}, \boldsymbol{\omega} \text{ s.t. } \big(z(0), v(0), \omega(0)\big) = (x, u, w),$$
$$z(k+1) = f\big(z(k), v(k), \omega(k)\big), \big(z(k), v(k)\big) \in Z, \omega(k) \in \mathbb{W}, \tag{6.9}$$
$$z(k) \in \overline{\mathcal{C}}_M^{\mathcal{K}} \cap \overline{\mathcal{R}}_M^{\mathcal{K}}, \ \forall k \in \mathbb{I}_{\geq 0} \Big\}.$$

We highlight that within $\overline{\mathcal{Z}}_M^{\mathcal{K}}$, the input sequence and the disturbance sequence can be chosen *dependent* on each other, and furthermore, $\overline{\mathcal{Z}}_M^{\mathcal{K}} \subseteq \overline{\mathcal{Z}}_{\mathcal{K}}^o$. Similar to the previous notation, we denote the projection of $\overline{\mathcal{Z}}_M^{\mathcal{K}}$ onto $\mathbb{X}$ by $\overline{\mathcal{X}}_M^{\mathcal{K}} := \{x \in \mathbb{X} : \exists u \in \mathbb{U}, w \in \mathbb{W} \text{ s.t. } (x, u, w) \in \overline{\mathcal{Z}}_M^{\mathcal{K}}\}$.

Theorem 6.5. *Suppose that system (2.7) is $\mathcal{K}$-robustly optimally operated at steady-state and let Assumptions 2.1–2.3 and 2.7 be satisfied. Then, for each $M \in \mathbb{I}_{\geq 1}$, system (2.7) is $\mathcal{K}$-robustly dissipative on $\overline{\mathcal{Z}}_M^{\mathcal{K}}$ with respect to the supply rate (6.6).*

The proof follows along the lines of the proof of Müller et al. (2015a, Theorem 3) and Theorem 5.20: We assume that the system is $\mathcal{K}$-robustly optimally operated at steady-state but not $\mathcal{K}$-robustly dissipative and lead this to a contradiction. To this end, one can associate $\mathcal{K}$-robust dissipativity with the respective available storage (compare to (4.20), (5.45) and Willems (1972))

$$S_a(x) := \sup_{\substack{T \geq 0, z(0) = x, \\ z(k+1) = f(z(k), v(k), \omega(k)), \\ (z(k), v(k), \omega(k)) \in \overline{\mathcal{Z}}_{\mathcal{K}}^o, \forall k \in \mathbb{I}_{\geq 0}}} \sum_{k=0}^{T-1} -\Big(s\big(z(k), v(k)\big) + \rho(w_{\max})\Big), \tag{6.10}$$

for which holds $S_a(x) < \infty$ for all $x \in \overline{\mathcal{X}}_M^{\mathcal{K}}$ if and only if the system is $\mathcal{K}$-robustly dissipative. From the assumption that the system is not $\mathcal{K}$-robustly dissipative, one constructs a cyclic state/input/disturbance sequence triple $(\boldsymbol{x}, \boldsymbol{u}, \boldsymbol{w})$ such that these sequences violate (6.4). This provides the desired contradiction.

We note again that in this setup it is sufficient to find, i.e., construct, *one* disturbance sequence for which we can derive the contradiction. This is in particular contrast to the approach in Section 4.2.3 and 5.3.1 where necessity of dissipativity is (and has to be) provided for all possible disturbance sequences.

As for the previous results on necessity of dissipativity, $\mathcal{K}$-robust dissipativity on $\overline{\mathbb{Z}}_{\mathcal{K}}^o$ is necessary and sufficient for $\mathcal{K}$-robust optimal operation at steady-state if $\overline{\mathbb{Z}}_M^{\mathcal{K}} = \overline{\mathbb{Z}}_{\mathcal{K}}^o$ for some $M \in \mathbb{I}_{\geq 1}$ (compare to Müller et al. (2015a, Corollary 1) as well as Section 4.2.3 and 5.3.1).

6.1.2 Linear systems

In the nominal and in the tube-based case (see Section 4.2.1), it is possible to show optimal operation at steady-state for linear systems when the cost function is (strictly) convex, the constraints are convex, and a Slater condition is satisfied (see, e.g., Damm et al. (2014); Diehl et al. (2011)). We want to investigate if an analogous result can be derived for our new notion as well. To this end, we consider the linear case, i.e., we reconsider the dynamic system

$$x(t+1) = Ax(t) + Bu(t) + w(t), \tag{6.11}$$

(compare to (2.20)). In the following discussion, we assume A to be asymptotically stable such that the mRPI set of the associated error dynamics is a feasible choice for $\Omega(w_{\max})$ (see Section 2.2.2). Moreover, the constraints as well as the disturbance set are considered convex and compact and, in addition, the disturbance set contains the origin in its interior. Hence, according to Diehl et al. (2011), the linear system is $\mathcal{K}$-robustly dissipative with respect to the supply rate $s(x, u) = \ell(x, v) - \ell(x_s(0), u_s(0))$ with a linear storage function for the nominal case ($w_{\max} = 0$). When it comes to $\mathcal{K}$-robust dissipativity, in the disturbed case we consider a linear storage function $\lambda(x) = \lambda^\top x$ and we need to show that

$$\lambda^\top(Ax + Bu + w) - \lambda^\top x \leq \ell(x, u) - \ell\Big(x_s(w_{\max}), u_s(w_{\max})\Big) + \rho(w_{\max}). \tag{6.12}$$

By linearity and by the guaranteed existence of a storage function for any undisturbed linear system, this boils down to proving the existence of a $\mathcal{K}_\infty$ function ρ satisfying

$$|\lambda| w_{\max} - \ell\Big(x_s(0), u_s(0)\Big) + \ell\Big(x_s(w_{\max}), u_s(w_{\max})\Big) \leq \rho(w_{\max}). \tag{6.13}$$

In order to be able to guarantee existence of such a ρ, $\ell(x_s(w_{\max}), u_s(w_{\max}))$ shall be continuous at $w_{\max} = 0$. If this is ensured, the left hand side of (6.13) is continuous at $w_{\max} = 0$ and a $\mathcal{K}_\infty$ function ρ always exists such that it satisfies the inequality (Rawlings and Mayne, 2012, Proposition 11).

Lemma 6.6. *Suppose that Assumptions 2.2, 2.3, and 2.7 are satisfied and that A is asymptotically stable. Let the optimal steady-state be determined by (6.1). Then, the cost evaluated at the optimal steady-state, $\ell(x_s(w_{\max}), u_s(w_{\max}))$, is continuous at $w_{\max} = 0$.*

By means of Lemma 6.6, we have guaranteed that—as expected—in the linear, strictly convex case, the system is always $\mathcal{K}$-robustly dissipative, and hence, $\mathcal{K}$-robustly optimally operated as steady-state.

6.1.3 Summary

By this new notion of $\mathcal{K}$-robust optimal steady-state operation, we introduced another possibility to investigate optimality—similar to the nominal case—only based on the dynamics, the stage cost function, and the constraints, i.e., independent of the MPC algorithm. The resulting statement can be interpreted as the steady-state operation being approximately optimal up to an error term depending on the largest disturbance. Even though the optimality statement is generally weaker than for the previous notions, it can provide some first insight into the system behavior and is appealing due to its independence of the underlying optimal control problem.

6.2 Comparison of optimality notions

Comparing the different notions of dissipativity and optimal operation at steady-state, one can see that the notion of robust dissipativity and $\mathcal{K}$-robust dissipativity are closely related to their nominal counterpart (compare Definition 2.25). While $\mathcal{K}$-robust optimality is a rather weak statement, robust optimality is much stronger providing information about the optimal *nominal* closed-loop trajectory (see Section 4.2.1). However, this is only possible due to considering the special structure of the underlying MPC algorithm. The notion of Ω-robust optimal steady-state operation is also particularly dependent on the specific structure of the underlying MPC algorithm. The additional degree of freedom provided by the free nominal initial state in Problem 4.1, which allows for better performance in general, makes it more difficult to prove steady-state optimality. As discussed in detail in Section 4.2, sequences of nominal states must be considered which do *not* satisfy the dynamics of the nominal system. If Ω-robust steady-state optimality can be shown, this is a strong statement, however, it is hard to satisfy. Considering stochastic optimal operation at steady-state, the additional stochastic information is directly considered within the optimality statement. However, for this notion it becomes significant that the derived statements are *a priori* statements, that is, we only consider information provided at initial time and the achieved closed-loop performance can differ significantly (compare Section 5.3).

6.2.1 Numerical example comparing dissipativity approaches

In order to compare the different notions of optimal operation at steady-state, we investigate them by means of an example. To this end, consider the linear two-dimensional system

$$x(t+1) = \begin{bmatrix} 0.8 & -0.6 \\ 0.6 & 0.8 \end{bmatrix} x(t) + \begin{bmatrix} 1 \\ 1 \end{bmatrix} u(t) + w(t), \tag{6.14}$$

with the constraints $\mathbb{X} = \{x \in \mathbb{R}^2 : \|x\|_\infty \leq 10\}$, $\mathbb{U} = \{u \in \mathbb{R} : |u| \leq 1\}$, and $\mathbb{W} = \{w \in \mathbb{R}^2 : \|w\|_\infty \leq 0.1\}$. Since the system has two eigenvalues on the boundary of the unit disc, we parametrize the input with the pre-stabilizing linear feedback $u = -Kx + v$, where $K = [0.54, 0.3]$, such that $v \in \mathbb{R}$ is the new manipulated input. An invariant outer approximation of the mRPI set Ω_∞ is computed by the algorithm presented in Raković et al. (2005), which results in a polytope with 50 vertices and non-empty interior. We consider the following quadratic stage cost function,

$$\ell(x,u) = (x_1 - a)^2 + (x_2 - a)^2 + u^2, \tag{6.15}$$

where $a \in \mathbb{R}$ with $|a| \leq 1$ is a parameter to be chosen subsequently in order to investigate the different notions of optimality. While for $a = 0$ this stage cost considers the tracking case, for $a \neq 0$ the optimum is not at a feasible steady-state of the system and, thus, this is an economic MPC problem.

Nominal dissipativity We start by investigating nominal steady-state optimality, that is, we set $w(t) \equiv 0$. Depending on a, the optimal steady-state is $x_s = \left(-\frac{a}{6}, \frac{a}{3}\right)^\top$ with $u_s = \frac{53}{300}a$. The system is dissipative with storage function $\lambda(x) = \left[\frac{19}{6}a, -\frac{17}{6}a\right]x + 60a$, and thus, it is optimally operated at steady-state. Note that dissipativity holds for any stabilizing K (with a suitably adapted storage function).

$\mathcal{K}$-robust dissipativity Here, we have $w_{\max} = 0.1\sqrt{2}$. Computing the optimal steady-state leads to the same steady-state as in the nominal setup. Using Lemma 6.6 and nominal steady-state optimality, we can see that this system is $\mathcal{K}$-robustly dissipative and $\mathcal{K}$-robustly optimally operated at steady-state as well. Again, optimality is independent of the choice of K.

Ω-robust dissipativity In the following, we investigate optimal operation with respect to the averaged case, however, similar results can be derived for the min-max case. For the averaged case, the optimal steady-state is again given by $\left(-\frac{a}{6}, \frac{a}{3}\right)^\top$. In case the nominal initial state is fixed, robust optimal operation follows directly from Proposition 4.9.

In case the nominal initial state is an optimization variable, we have to investigate Ω-robust dissipativity, i.e., one has to find a storage function λ such that

$$\max_{x \in (\{f(z, v - Kz, 0)\} \oplus 2\Omega_\infty) \cap \overline{\mathbb{X}}} \lambda(x) - \lambda(z) \leq \ell^{\text{int}}(z, v) - \ell^{\text{int}}(z_s^{\text{int}}, v_s^{\text{int}}), \tag{6.16}$$

for all $(z, v) \in \overline{\mathbb{Z}}_{\text{init}}^o$. Note that ℓ^{int} is strictly convex and has its minimum at $x = (a, a)$, $u = K[a, a]^\top$. If $a = 0$, we can choose $\lambda = 0$ to satisfy (6.16). In case of $a \neq 0$, there always exist state and input combinations $(\tilde{z}, \tilde{v}) \in \overline{\mathbb{Z}}$ in the neighborhood of the robustly optimal steady-state $(z_s^{\text{int}}, v_s^{\text{int}})$ such that $\tilde{z} \in (\{f(\tilde{z}, \tilde{v} - K\tilde{z}, 0)\} \oplus 2\Omega_\infty) \cap \overline{\mathbb{X}}$ and $\ell^{\text{int}}(\tilde{z}, \tilde{v}) - \ell^{\text{int}}(z_s^{\text{int}}, v_s^{\text{int}}) < 0$. For these states, the left hand side of (6.16) is always greater than or equal to zero, while the right hand side is strictly smaller than zero. Hence, the system is not Ω-robustly dissipative in this case. For $a = 0$, the system is Ω-robustly optimally operated at steady-state (independent of the choice of K).

Stochastic dissipativity Now, we assume the disturbance $\mathbb{W}$ to be equipped with a uniform distribution. Proving stochastic dissipativity according to Definition 5.13 can be derived for $a = 0$ by choosing $\lambda = -x^\top P x + c$, where $P \succ 0$ follows from the associated algebraic Riccati equation. In case $a \neq 0$, one can choose an ansatz of the form $\lambda(x) = -x^\top \bar{P} x + \bar{p}^\top x + \bar{c}$, with $\bar{P} \succ 0$, and derive a linear matrix inequality providing stochastic dissipativity, however, in this case $\bar{P}$ is usually different to the solution of the algebraic Riccati equation. Thus, the system is stochastically optimally operated at steady-state for all $|a| \leq 1$ (and for arbitrary stabilizing K).

Concluding discussion Concluding this example, one can show optimal operation at steady-state based on comparison functions, where only knowledge of $w_{\max}$ is needed. With the additional information about stochastic distribution, one can show stochastic optimal operation. For the notion based on nominal systems, Ω-robustly optimal operation at steady-state is achieved only in the tracking case $a = 0$. For $a \neq 0$, the system is *not* Ω-robust optimally operated at steady-state. As discussed above, this is in general a stricter notion since also nominal sequences which are not solutions of the nominal system (2.1) are considered, and the example does not satisfy this stricter notion for $a \neq 0$.

6.3 Numerical example

In this section, we apply the presented approaches for handling disturbances in an economic MPC framework to a numerical example. We restrict ourselves to a linear example to be able to consider all approaches presented. Additionally, we apply three standard approaches from robust stabilizing MPC to this problem in order to highlight that these can be disadvantageous in the economic setup.

The example considered is the linearized model of a continuous stirred tank reactor (CSTR). The original (nonlinear and continuous-time) model is introduced in Henson and Seborg (1997). It describes an exothermic, irreversible reaction, $A \to B$, under the assumption of a constant liquid volume. The model consists of two states, the concentration c^A of the reactant A and the temperature T in the reactor. The control input is the cooling temperature T^c. The feed flow concentration as well as the feed flow temperature can be seen as disturbances. In Pannocchia and Kerrigan (2005), this model is used for stabilizing tracking MPC. Thus, the origin is shifted to the desired unstable steady-state $(c_s^A, T_s) = (0.5 \, \text{mol}/\text{L}, 350 \, \text{K})$ with $T_s^c = 300 \, \text{K}$. Using a sampling time of $6 \, \text{s}$ and introducing deviation variables, the following linearized discrete-time model is derived in Pannocchia and Kerrigan (2005):

$$x(t+1) = \begin{bmatrix} 0.7776 & -0.0045 \\ 26.6185 & 1.8555 \end{bmatrix} x(t) + \begin{bmatrix} -0.0004 \\ 0.2907 \end{bmatrix} u(t) + \begin{bmatrix} -0.0002 & 0.0893 \\ 0.1390 & 1.2267 \end{bmatrix} w(t). \quad (6.17)$$

We note that x_1 and x_2 represent the deviations from c_s^A and T_s, respectively. Similarly, the deviation from T_s^c is denoted by u.

With regard to the constraints, we consider the same bounds as proposed in Pannocchia and Kerrigan (2005). The state is restricted to

$$\mathbb{X} = \Big\{ x \in \mathbb{R}^2 : |x_1| \leq 0.5, |x_2| \leq 5 \Big\} \quad (6.18)$$

and the input to

$$\mathbb{U} = \{ u \in \mathbb{R} : |u| \leq 15 \}. \quad (6.19)$$

The disturbance is bounded and *uniformly* distributed over the set

$$\mathbb{W} = \Big\{ w \in \mathbb{R}^2 : |w_1| \leq 2, |w_2| \leq 0.1 \Big\}. \quad (6.20)$$

As system (6.17) is unstable, a linear error feedback of the form $u = v + K(x - z)$ (compare to (2.21)) is employed, where, as in the previous chapters, z denotes the states and v the input of the nominal system, respectively. We choose $K = [-0.6457 \; -5.4157]$. An

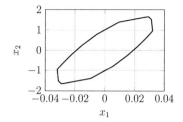

Figure 6.1: Invariant outer approximation of the mRPI set for the CSTR example considered.

invariant outer approximation of the mRPI set Ω_∞ for the associated error dynamics (2.22) (see Section 2.2.2) is computed by the algorithm presented in Raković et al. (2005). The resulting set is a polytope with 54 vertices depicted in Figure 6.1.

While in Pannocchia and Kerrigan (2005), the goal is to stabilize the desired steady-state, we consider a different objective. Namely, we are interested in minimizing the asymptotic average of the economic stage cost function

$$\ell(x, u) = x_1 + \begin{cases} 10(x_2 + 2)^2 & \text{for } x_2 < -2, \\ 0 & \text{for } -2 \leq x_2 < 2, \\ 10(x_2 - 2)^2 & \text{for } 2 \leq x_2. \end{cases} \qquad (6.21)$$

The first term of this cost function accounts for minimizing the concentration of the reactant A, which corresponds to a maximization of the concentration of the desired product B. The second term can be interpreted as a soft constraint on the temperature, which is desired to be kept in the interval $x_2 \in [-2, 2]$, whereas the hard constraints guarantee $x_2 \in [-5, 5]$. Furthermore, we note that this stage cost function is convex and independent of the input.

For the subsequent results, 100 simulations were run, all starting at the origin. The disturbance sequences considered are drawn from $\mathbb{W}$ with uniform distribution, and each approach is run with the *same* 100 realizations of the disturbance sequences. For all approaches the prediction horizon is fixed to $N = 20$ and we simulate over 200 closed-loop iterations. It will later become apparent that this is sufficient for providing convergence to the asymptotic average performance. The performance results provided are determined by summing up the stage cost function for all 200 closed-loop states for each of the 100 realizations and taking the average thereof.

In the next sections, we start by applying standard approaches from robust stabilizing MPC to the above example, followed by the min-max and the tube-based approach. At the end, we apply the approaches taking stochastic information into account. An overview over all the considered approaches is provided in Table 6.1.

6.3.1 Robust stabilizing MPC

In this section, three approaches from robust stabilizing MPC are directly applied to the economic CSTR example. All three approaches consider an objective that is independent

Table 6.1: Comparison of the different algorithms presented in Section 6.3. The objective is stated with respect to the nominal system (z, v) or the real system (x, u), respectively, depending on the considered setup. The constraint tightening is either constant (C), i.e., with respect to the mRPI set Ω_∞, or with respect to a tightening which varies with the prediction time (T), see, e.g., Section 2.2.2. The performance bounds are the numerical results for the average performance for the CSTR example considered.

Algorithm	Objective	Nominal initial state	Constraint tightening	Stochastic information	Performance result for numerical example
Section 6.3.1: Robust stabilizing MPC					
Algorithm 2.16	$\sum_{k=0}^{N-1} \ell(z(k), v(k)) + V_f(z(N))$	Free	C	-	13.7158
Algorithm 2.16 fixed IC	—"—	Fixed	C	-	0.4640
Algorithm 5.4 w/(6.22)	—"—	Fixed	T	-	0.1222
Section 6.3.2: Min-max robust economic MPC					
Algorithm 3.8	$\sum_{k=0}^{N-1} \max_{\omega \in \Omega_\infty} \ell(z(k)+\omega, v(k)+K\omega) + \overline{V}_f(z(N))$	Free	C	-	-0.0227
Algorithm 3.8 fixed IC	—"—	Fixed	C	-	-0.0064
Problem 3.12	$\sum_{k=0}^{N-1} \max_{\omega \in \Omega_k} \ell(z(k)+\omega, v(k)+K\omega) + \tilde{V}_f(z(N))$	Fixed	T	-	-0.0275
Section 6.3.3: Tube-based robust economic MPC					
Algorithm 4.2	$\sum_{k=0}^{N-1} \ell^{\text{int}}(z(k), v(k)) + V_f^{\text{int}}(z(N))$	Free	C	-	-0.0283
Algorithm 4.8	—"—	Fixed	C	-	-0.0091
Section 6.3.4: Stochastic information					
Algorithm 5.4	$\sum_{k=0}^{N-1} \ell_k^{\text{int}}(z(k), v(k)) + \overline{V}_f^{\text{int}}(z(N))$	Fixed	T	x	-0.0290
Algorithm 5.23	$\sum_{k=0}^{N-1} \ell_\infty^{\text{int}}(z(k), v(k)) + V_{f,\infty}^{\text{int}}(z(N))$	Fixed	C	x	-0.0209
Section 6.3.5: Classical min-max approach $(N = 5)$					
Algorithm 3.2	$\max_{w \in W^N} \sum_{k=0}^{N-1} \ell(x(k), u(k)) + V_f^c(x(N))$	Fixed	T	-	-0.0289
Algorithm 3.8	$\sum_{k=0}^{N-1} \max_{\omega \in \Omega_\infty} \ell(z(k)+\omega, v(k)+K\omega) + \overline{V}_f(z(N))$	Free	C	-	-0.0230
Algorithm 3.8 fixed IC	—"—	Fixed	C	-	-0.0053
Problem 3.12	$\sum_{k=0}^{N-1} \max_{\omega \in \Omega_k} \ell(z(k)+\omega, v(k)+K\omega) + \tilde{V}_f(z(N))$	Fixed	T	-	-0.0213

from the influence of the disturbance

$$J_N\Big(z(0|t), \boldsymbol{v}(t)\Big) = \sum_{k=0}^{N-1} \ell\Big(z(k|t), v(k|t)\Big) + V_f\Big(z(N|t)\Big) \qquad (6.22)$$

(compare to (2.18)). As we will see later, this results in a poor asymptotic average performance for this example. Two approaches are based on Algorithm 2.16. The first approach considers Problem 2.15. In particular, this problem considers a constant tightening with respect to the mRPI set Ω_∞ and allows for the nominal initial state as an optimization variable. The second approach fixes the nominal initial state to follow a trajectory (compare to Section 4.2.1), that is, we replace (2.17b) for all $t \in \mathbb{I}_{\geq 1}$ by $z(0|t) = z^*(1|t-1)$ and $z(0|0) = x(0)$ (Algorithm 2.16 fixed IC). For the terminal controller, we use a linear controller with the same gain K as for the error feedback. The terminal region is chosen to be the maximal output admissible set (see, e.g., Mayne et al. (2005)) and a quadratic terminal cost is derived based on the approach presented in Amrit et al. (2011)[1].

The third approach considered in this section is based on the stabilizing approach in Chisci et al. (2001), which has briefly been reviewed in Section 2.2.2. The problem and the algorithm considered resemble Problem 5.3 and Algorithm 5.4, respectively. For this setup, a constraint tightening is taken into account that varies with prediction time allowing to consider the growing error sets. In contrast to Problem 5.3, the nominal counterpart (6.22) is employed as the objective (Algorithm 5.4 with (6.22)). Again, a linear controller with gain K is used as the terminal controller. The terminal region is determined according to Chisci et al. (2001), the terminal cost follows according to the methodology presented in Amrit et al. (2011).

The nominal optimal steady-state for these approaches lies at $x_s = (-0.0358, 2.0009)^\top$ with $u_s = -2.6113$. For these approaches, we cannot determine any a priori bounds on the performance. When running the simulations for the above mentioned 100 realizations we receive the following asymptotic average performance results:

Algorithm 2.16	Algorithm 2.16 fixed IC	Algorithm 5.4 with (6.22)
13.7158	0.4640	0.1222

In Figure 6.2, one can see the convex hull of all closed-loop states (after a settling time of 10 closed-loop iterations steps plotted over the contour lines of the stage cost function (6.21)). It is interesting to note the differences and the similarities in the behavior of the approaches. Since the stage cost function considered in all of the approaches is independent of the influence of the disturbance, the optimal steady-state is at the edge of the soft constraint on the temperature ($x_2 \leq 2$) while minimizing the concentration (x_1). The two approaches based on Algorithm 2.16 fixed IC (dash-dotted) and Algorithm 5.4 (dotted), show a similar behavior. Both approaches keep the closed-loop state in the vicinity of the optimal steady-state, but the closed-loop system is affected by disturbances, and hence, it gets deviated and violates the soft constraint on the temperature imposed by the stage cost function. However, since the second algorithm uses direct state information in contrast to the first approach, which is independent of state measurements, it achieves a better performance.

---

[1]Note that the cost considered here is not $\mathcal{C}^2$. However, we can still find a quadratic upper bound for the exact cost on the terminal region such that the procedure presented in Amrit et al. (2011) can be applied.

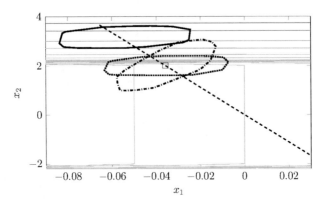

Figure 6.2: Contour plot of the stage cost ℓ in (6.21). All feasible steady-states are given by the dashed line. The convex hull of the closed-loop states after a settling time of 10 closed-loop iterations are depicted by the sets (Algorithm 2.16 – solid; Algorithm 2.16 fixed IC – dash-dotted; Algorithm 5.4 with objective (6.22) – dotted). The nominal optimal steady-state x_s is depicted by the square.

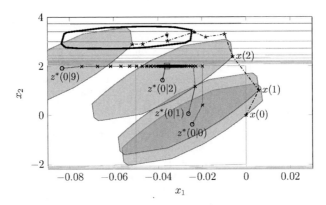

Figure 6.3: Contour plot of the stage cost ℓ in (6.21). The convex hull of the closed-loop states of Algorithm 2.16 after a settling time of 10 closed-loop iterations are depicted by the solid set, the first ten iterations by the dash-dotted line and the stars. The nominal open-loop trajectories for iteration 0, 1, 2, and 9 are provided by the dotted lines and the crosses, their initial states by the circles. In gray are the mRPI sets of the same iterations centered at the initial states of the open-loop trajectories. The optimal steady-state x_s is depicted by the square.

In order to understand the poor behavior of the first approach based on Algorithm 2.16 (solid), additional information is provided in Figure 6.3. When considering Problem 2.15, one can see that the nominal initial state is an optimization variable of the underlying optimal control problem, which has to satisfy $x(t) \in \{z(0|t)\} \oplus \Omega_\infty$ (gray set). The optimization places the nominal initial state (circles) as far into negative direction of the concentration as possible, ignoring the behavior of the real system. However for this particular example, it turns out that the resulting closed-loop trajectory (stars) is driven such that it violates the soft constraint on the temperature and thus, results in a poor asymptotic average performance.

Summarizing, this first part of the example highlights why just transferring approaches from robust stabilizing MPC to the economic setup might not lead to a desirable asymptotic average performance of the closed-loop system.

6.3.2 Min-max robust economic MPC

In this section, we present the results for the three min-max robust economic approaches based on tubes presented in Section 3.2. We compare two approaches based on Algorithm 3.8, where the first approach employs Problem 3.6 (Algorithm 3.8) and the second one uses a fixed nominal initial condition (Algorithm 3.8 fixed IC) to recast a trajectory of the nominal system according to the discussion in Remark 3.10. The third approach considers growing sets representing the error evolution along the prediction (Problem 3.12). The terminal cost functions and the terminal regions are computed according to the discussions in Section 3.2.1 and 3.2.2, respectively. Considering the robust optimal steady-state, it follows $z_s^{\max} = (-0.0066, 0.3672)^\top$ with $v_s^{\max} = -0.4793$.

For the relatively long prediction horizon of $N = 20$, the classical min-max approach could not be solved due to the underlying computational complexity. Thus, results are only stated for the tube-based approaches. We provide a comparison for this approach with the min-max approaches based on tubes subsequently with a shorter horizon in Section 6.3.5.

Considering the a priori bounds on the asymptotic average performance in (3.20), we can determine $\max_{\omega \in \Omega_\infty} \ell(z_s^{\max} + \omega, K\omega + v_s^{\max}) = 0.0254$ which provides the bound for the first two approaches. For the third approach, the bound in (3.42) is $\max_{\omega \in \Omega_N} \ell(z_s^{\max} + \omega, K\omega + v_s^{\max}) + \max_{w \in W} \frac{1+\bar{\varepsilon}}{2}(A_{\text{cl}}^N w)^\top P_\varepsilon A_{\text{cl}}^N w + p^\top A_{\text{cl}}^N w = 0.0255.$[2] As mentioned before, these bounds are conservative since they consider the worst case in the mRPI set. Nevertheless, these bounds are already better than the achieved performance results in the previous section. The mentioned conservatism becomes obvious when comparing the bounds to the simulated results for the asymptotic average performance:

Algorithm 3.8	Algorithm 3.8 fixed IC	Problem 3.12
−0.0227	−0.0064	−0.0275

In Figure 6.4, the convex hulls of the respective closed-loop trajectories are shown. Taking these into account, the relatively better performance of the three approaches compared to the nominal approaches from Section 6.3.1 becomes obvious. By considering the worst case within the stage cost function in Problem 3.6, the system is operated such that all states in the mRPI set satisfy the soft constraint. Considering Algorithm 3.8 with fixed

---

[2]Choosing $\varepsilon = 10.17$, we determine $\bar{\varepsilon} = 4835.8$.

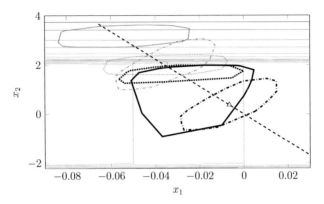

Figure 6.4: Contour plot of the stage cost ℓ in (6.21). All feasible steady-states are given by the dashed line. The convex hull of the closed-loop states after a settling time of 10 closed-loop iterations are depicted by the sets (Algorithm 3.8 – solid; Algorithm 3.8 fixed IC – dash-dotted; Problem 3.12 – dotted). The robust optimal steady-state z_s^{max} is depicted by the triangle. The graphs from Figure 6.2 are restated in gray.

initial condition, the system can be shown to be robustly optimally operated at steady-state (see Section 4.2.4). Since this setup is independent of state measurements (except for the error feedback), the nominal system converges to the optimal steady-state and stays there for all times. Thus, the closed-loop states of the real system remain in a vicinity of the optimal steady state z_s^{max} namely in the mRPI set centered at z_s^{max}. Since the soft constraint is never violated, the asymptotic average performance is significantly better than for the approaches presented in Section 6.3.1. For Algorithm 3.8, which is based on Problem 3.6, direct information about the actual state is considered within the underlying optimal control problem, and thus, the system can be operated closer to the soft constraint without violating it. This leads to a significant improvement of the asymptotic average performance. Taking the growing error sets (3.31) instead of the mRPI set into account, as is done in Problem 3.12, clearly reduces conservatism and thus improves the asymptotic average performance even further.

6.3.3 Tube-based robust economic MPC

The third class of approaches that we consider for the particular example are the tube-based robust economic MPC approaches presented in Chapter 4. These approaches are based on taking the average over the mRPI set Ω_∞ into account within the stage cost function. As in the previous section, a difference can be made in the way the nominal initial state is handled. While the first algorithm keeps the nominal initial state as an additional optimization variable (Algorithm 4.2), the second algorithm fixes the nominal initial state to follow a trajectory (Algorithm 4.8). Again, this results in an algorithm being independent of the measured state. Computing the robust optimal steady-state results in $z_s^{\mathrm{int}} = (-0.0094, 0.5254)^\top$ with $v_s^{\mathrm{int}} = -0.6857$. As for the nominal case, the terminal region

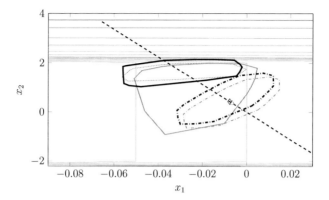

Figure 6.5: Contour plot of the stage cost ℓ in (6.21). All feasible steady-states are given by the dashed line. The convex hull of the closed-loop states after a settling time of 10 closed-loop iterations are depicted by the sets (Algorithm 4.2 – solid; Algorithm 4.8 – dash-dotted). The robust optimal steady-state z_s^{int} is depicted by the diamond. The graphs from Figure 6.4 are restated in gray.

is chosen to be the maximal output admissible set and a quadratic terminal cost is derived based on the approach presented in Amrit et al. (2011) with respect to the integrated stage cost function ℓ^{int}.

Computing the performance bound in (4.7), we end up with $\ell^{\text{int}}(z_s^{\text{int}}, v_s^{\text{int}}) = -0.00087$. We recall that this bound has to be understood as a bound on the average performance result for the real closed-loop system, averaged over all possible disturbances by integrating over the mRPI set. Hence, this is *not* a bound on the asymptotic average performance of the closed loop. The asymptotic average performances result in:

Algorithm 4.2	Algorithm 4.8
−0.0283	−0.0091

In Figure 6.5, the respective convex hulls for the closed-loop trajectories are provided. As is the case for the min-max robust economic approach, fixing the nominal initial state (Algorithm 4.8), and thus, ignoring new information about the current state, results in a worse asymptotic average performance. According to Proposition 4.9, one can see that for Algorithm 4.8, the system is robustly optimally operated at steady-state.

Comparing the tube-based approaches and those considering the min-max concept based on tubes to each other, that is, comparing Algorithm 4.2 to Algorithm 3.8 and comparing Algorithm 4.8 to Algorithm 3.8 fixed IC, a general trend can be identified: The approaches based on averaging are shifted along the line of steady-states towards the soft constraint on the temperature. This is due to the averaging approaches allowing for some states within the mRPI sets centered at the nominal trajectory to violate the soft constraint. The higher cost induced by those states can be balanced by considering the average over all states within the mRPI set. In contrast, the min-max approaches only consider the worst case within the mRPI set. This leads to a prevention of violating the soft constraint with any

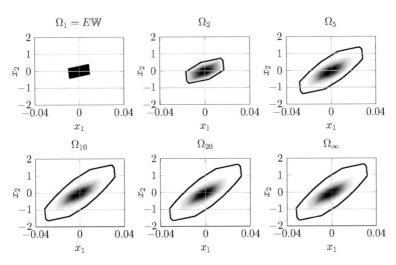

Figure 6.6: Error sets Ω_k for different prediction steps $k = \{1, 2, 5, 10, 20\}$ and for the mRPI set Ω_∞ for the numerical example. The shading represents the associated PDFs $\rho_{\Omega_k}(\epsilon)$. A darker color stands for a higher probability.

state within the mRPI set.

6.3.4 Stochastic information

The last class of approaches that we apply to the numerical example considers stochastic information about the disturbance. Two different approaches are taken into account. The first approach optimizes for the expected value of the cost along the prediction. In particular, the constraints are tightened with respect to the growing error set and, in addition, the distribution over each of these sets is considered (Algorithm 5.4). The second approach is similar to the previously considered averaging approach, however, it takes the distribution over the mRPI set within the predictions into account (Algorithm 5.23). It considers a tightening that is only dependent on the mRPI set and not prediction-time varying.

In Figure 6.6, the error sets for different prediction times as well as their distributions are depicted. These are computed by gridding the disturbance set and using a discrete convolution. The distribution over the mRPI set is computed by repeating the convolution up to a predefined relative error ($\varepsilon_{\mathrm{err}} = 10^{-4}$) of two consecutive iterations; in this case 49 iterations are needed. However for this example, after 10 iterations we already have a good approximation of the distribution over the mRPI set Ω_∞.

The stochastic optimal steady-state for these approaches is at $z_s^\infty = (-0.0227, 1.2692)^\top$ with $v_s^\infty = -1.6565$. Computing the bounds on the expected asymptotic average performance, we can see that $\ell_\infty^{\mathrm{int}}(z_s^\infty, v_s^\infty) = -0.0243$ (see Theorem 5.5), whereas for the quadratic approximation in (5.34), we receive $\ell(z_s^\infty, v_s^\infty) + \frac{1}{2} \int_{\mathbb{W}} \epsilon^\top P \epsilon \rho_{\mathbb{W}}(\epsilon) d\epsilon = 0.2261$. The difference between the two bounds is quite large for this example, which is caused by the rather large terminal region and the inevitable symmetry of the terminal cost (see

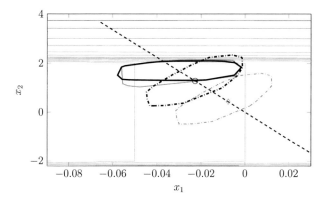

Figure 6.7: Contour plot of the stage cost ℓ in (6.21). All feasible steady-states are given by the dashed line. The convex hull of the closed-loop states after a settling time of 10 closed-loop iterations are depicted by the sets (Algorithm 5.4 – solid; Algorithm 5.23 – dash-dotted). The stochastic optimal steady-state z_s^∞ is depicted by the circle. The graphs from Figure 6.5 are restated in gray.

Section 5.2). Due to the quadratic approximation of the terminal cost, the predictable a priori asymptotic average performance bound (5.34) is quite conservative, compared to the closed-loop performance achieved in simulation, see below. As introduced in Section 5.2.1, we can determine a better a priori performance bound by means of using a different control horizon (N) and prediction horizon ($N + \tilde{N}$) within the terminal cost. The basic idea for finding P is the same as before, however, the set over which the quadratic terminal cost must be satisfied is smaller. This can be seen when comparing the performance bound for different values of $\tilde{N}$:

$\tilde{N}$	1	3	5	9
Perf. Bound	0.2255	0.2201	0.1841	0.1418

We note that these bounds are still larger than the bound $\ell_\infty^{\text{int}}(z_s^\infty, v_s^\infty)$ since a quadratic approximation of the terminal cost is used in all cases. As for the tube-based robust economic setup, this is *not* a bound on the asymptotic average performance of the closed loop but must be understood in the averaging sense (see discussion after Theorem 5.5).

The terminal region for the optimal control problem underlying Algorithm 5.23 is chosen to be the maximal output admissible set, the terminal cost is derived by means of the approach presented in Amrit et al. (2011) with respect to the stage cost function ℓ_∞^{int}.

The asymptotic average performance is received as follows:

Algorithm 5.4	Algorithm 5.23
−0.0290	−0.0209

In Figure 6.7, the respective hulls of the closed-loop states are depicted. It turns out that the closed-loop state resulting from Algorithm 5.4 is close to the bound of the soft constraint while trying to reduce the concentration c_A as much as possible. Due to the stage cost function being dependent on the time-varying tubes Ω_k, the algorithm does not violate the soft constraint, which would result in a poor performance.

Considering the performance for the closed loop resulting from Algorithm 5.23, we have to take also Figure 6.6 into account. One can see that the states towards the edge of the mRPI set are unlikely to occur, which depends on the uniform distribution of the disturbance. Thus, the closed loop (as well as the robust optimal steady-state) is closer to the soft constraint ($x_2 \leq 2$) than for the other approaches based on nominal information only.

In Figure 6.8, the transient average performance

$$J_T(\boldsymbol{x}, \boldsymbol{u}) = \frac{1}{T} \sum_{t=0}^{T-1} \ell\Big(x(t), u(t)\Big) \tag{6.23}$$

is depicted for growing T and for 100 simulations. In order to provide a clear presentation, we only present the results for representative MPC approaches, namely Algorithm 2.16 fixed IC, Algorithm 4.2, Algorithm 5.4, and Algorithm 5.23. One can see that the statistical mean values of the 100 simulations (dash-dotted lines) converge to the asymptotic average performance (dotted lines) as T grows, and already after 80 closed-loop iterations, the asymptotic average performance is converged.

6.3.5 Results for classical min-max approach

In this section, we consider additionally the classical min-max approach (see Section 3.1). Due to the computational complexity, we were not able to simulate the classical approach with $N = 20$, but only with $N = 5$. In order to make the other approaches comparable, we present the result for all other min-max approaches with the same, shorter horizon and apply the same linear input parametrization (compare Remark 3.4). For this example, we consider a smaller number of simulations, namely only 20 simulations with 200 closed-loop iterations each. Again, the same disturbance sequences are applied to all approaches.

The optimal steady-states of the classical min-max approach is equivalent to the optimal steady-state in the nominal approach, see Section 6.3.1. Thus, it follows that $x_s^c = (-0.0358, 2.0009)^\top$ and $c_s^c = 8.2019$ noting that due to the parametrization $u_s = Kx_s^c + c_s^c$. Considering the performance bound for the classical min-max approaches, one can use an idea similar to Sontag and Wang (1995, Remark 2.4). This leads to the quadratic terminal cost with $P = \begin{bmatrix} 2753.14 & -59.14 \\ -59.14 & 2.04 \end{bmatrix}$ and $p = [3.2473, -0.0105]^\top$ as well as to $\rho(\|w\|_{\Phi^2}) = 0.1431\|w\|_{\Phi^2}^2 + 0.0277\|w\|_{\Phi^2}$, with $\Phi = \begin{bmatrix} 0.5 & 0 \\ 0 & 10 \end{bmatrix}$ resulting in a performance bound of $\phi(w_{\max}) = 0.3258$.

The bounds on the performance for the other approaches are derived as in Section 6.3.2. They are given by $\max_{w \in \Omega_\infty} \ell(z_s^{\max} + \omega, K\omega + v_s^{\max}) = 0.0254$ for the two approaches based on Algorithm 3.8 and $\max_{w \in \Omega_N} \ell(z_s^{\max} + \omega, K\omega + v_s^{\max}) + \max_{w \in W} \frac{1+\bar{\varepsilon}}{2}(A_{cl}^N w)^\top P_\varepsilon A_{cl}^N w + p^\top A_{cl}^N w = 571.31$ for the approach based on Problem 3.12.[3] The last bound is really conservative. While the first term (for $N = 5$) results in $\max_{w \in \Omega_N} \ell(z_s^{\max} + \omega, K\omega + v_s^{\max}) = 0.0218$, the second term is 571.28. This is due to the relatively short prediction horizon and the conservative quadratic over-approximation of the terminal cost function in (3.42).

As we can see, these are conservative bounds for the asymptotic average performance of the closed loop:

---

[3]Choosing $\varepsilon = 6.45$, we determine $\bar{\varepsilon} = 4787.8$.

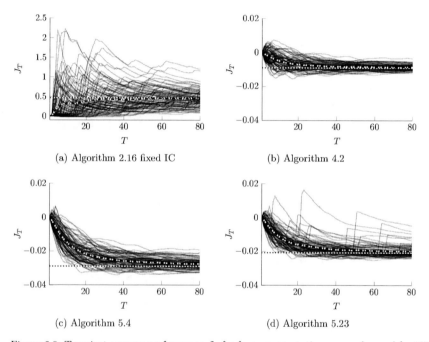

(a) Algorithm 2.16 fixed IC

(b) Algorithm 4.2

(c) Algorithm 5.4

(d) Algorithm 5.23

Figure 6.8: Transient average performance J_T for four representative approaches and for 100 simulations each (solid lines). The dotted lines depict the asymptotic average performances, the dash-dotted lines represent the statistical mean value of the 100 simulations. We note the different scaling of the J_T-axis in (a).

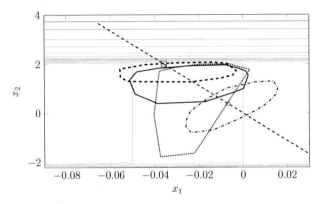

Figure 6.9: Contour plot of the stage cost ℓ in (6.21). All feasible steady-states are given by the dashed line. The convex hull of the closed-loop states after a settling time of 10 closed-loop iterations are depicted by the sets (Algorithm 3.2 – dashed; Algorithm 3.8 – solid; Algorithm 3.8 fixed IC – dash-dotted; Problem 3.12 – dotted). The optimal steady-states $z_s^{\max}$ and x_s^c are depicted by the triangle and the square, respectively.

Algorithm 3.2	Algorithm 3.8	Algorithm 3.8 fixed IC	Problem 3.12
-0.0289	-0.0230	-0.0053	-0.0213

As expected, the classical approach provides the best asymptotic average performance. While the closed-loop trajectories of all approaches stay within the desired temperature range, the classical approach gets closest to this soft constraint (see Figure 6.9). With this shorter horizon, it turns out that the min-max approach based on Problem 3.12, that is, the approach taking a tightening depending on the prediction step into account is worse than Algorithm 3.8. However when investigating the simulations, this is due to the short prediction horizon, which is also indicated by the different shape of the convex hulls of the closed-loop states in Figure 6.4 and Figure 6.9.

6.3.6 Concluding discussion

Summing up the observations made in this numerical example, we can see that more information provides a better asymptotic average performance result. This can be seen in the stochastic case providing the best result. Additionally, a time-varying tightening, which can as well be considered as taking more information into account, can significantly improve the asymptotic average performance, as becomes apparent in the min-max case in Section 6.3.5. However, the computational complexity of the classical min-max approach obviates its application. Comparing the three tube-based approaches, we can see that the performance result improves with an increasing amount of information. For all approaches based on the min-max idea, it is striking how much the actual asymptotic average performance deviates from the derived bounds highlighting that the bounds (i) have to hold for all possible disturbance sequences and (ii) are deteriorated by using approximate quadratic terminal cost functions. For the other approaches, the bounds on the performance do not provide statements for the performance of the closed loop, and thus, have to be interpreted differently according to their respective definitions.

6.4 Summary

In this chapter, we introduced a new notion of optimal operation at steady-state which is—inspired by the nominal notion—only depending on the system dynamics, the stage cost function, and the constraints. Subsequently, the different notions of optimal operation at steady-state were compared to each other, in discussion as well as with a numerical example. Moreover, we presented a detailed numerical example applying and comparing all the introduced robust economic MPC approaches. We investigated all the determined performance bounds and provided an elaborate discussion on the resulting closed-loop behavior of the different algorithms presented.

Chapter 7

Conclusions

In this chapter, we summarize the main results of the thesis. Additionally, we highlight possible directions for future research.

7.1 Summary

In this thesis, we presented concepts on how to incorporate information on disturbances within an economic MPC framework. To this end, we introduced three concepts with several algorithmic setups to approach this idea. A thorough theoretical analysis was performed for all of the approaches. We connected conceptual ideas form robust stabilizing MPC and nominal economic MPC in order to provide a framework for robust economic MPC.

In Chapter 3, the idea of min-max robust control, which is often used in the framework of robust MPC, was applied to the economic setup. Two main approaches were presented. While the first approach considers the worst-case disturbance sequence along the open-loop predictions within the objective, the second approach approximates the influence of the disturbance using ideas from tube-based robust MPC. By this, the worst case of the disturbance is not evaluated for open-loop predictions of the disturbance sequence, but considered separately at each prediction step.

In Chapter 4, we picked up on the second idea but replaced the worst-case representation by averaging over all possible real system states induced by the disturbance acting on the system. The state information is provided by means of invariant error sets. However, instead of the worst case, the average performance over these sets is considered within the objective function. This can be understood as assuming a uniform distribution over the invariant error sets. The structure of the considered MPC algorithm played a crucial step in the analysis.

In Chapter 5, additional information on the disturbances was exploited by considering distributions of the disturbance. Two approaches were presented for linear systems, and both are based on predicting the propagation of the distribution. The first approach explicitly computes the propagation of the disturbance along the prediction horizon and incorporates the expected value in the objective. The second approach makes use of the converged distribution of the disturbance and provides a setup comparable to the one in Chapter 4, but with additional distribution information replacing the (probably) poor estimate of a uniform distribution over the associated invariant error set.

For all of the considered setups in Chapter 3, 4, and 5, bounds on the asymptotic average performance were derived. In addition, constructive methods to compute terminal cost functions were provided, which is crucial for proving the derived performance results as well as for practical implementation. Moreover, we investigated optimal operation at steady-state for all of the presented approaches. As is known from literature, this is an interesting property for economic MPC schemes since it provides approximate a priori information about the optimal closed-loop behavior. However, in robust MPC, these results are only approximate in the sense that they are open-loop statements, i.e., they provide information for all possible disturbance realizations, whereas the online MPC algorithms operates in closed-loop and usually reacts to the disturbances. Since these results on optimal operation at steady-state were based on appropriate dissipativity notions, we furthermore investigated converse results on the necessity of dissipativity for all of the notions on steady-state optimality, which led to the result that the notions of dissipativity introduced are (under certain assumptions) precise, i.e., not only conservative sufficient conditions, for optimal operation at steady-state. For the approaches in Chapter 3 and 4, we also derived results on asymptotic stability. These results are, again, based on particular notions of (strict) dissipativity. Asymptotic stability is of interest in robust economic MPC, as it shows that there does not only exist a robust optimal steady-state, but this optimal steady-state is asymptotically stable for the system under the proposed robust economic MPC feedback strategies.

In Chapter 6, a notion of optimal steady-state operation was introduced which is only based on the stage cost function, the dynamics, and the constraints. This is explicitly independent of the underlying optimal control problem in the respective MPC algorithms. Furthermore, we compared the different optimality notions as well as the different MPC schemes presented in this thesis to each other and discussed the differences in detail, in part by employing numerical examples. Considering the performance results for the presented numerical example of a (simplified) continuous-stirred tank reactor, it turned out that considering "more information", for example by means of stochastic information but also by time-variant constraint tightening, might improve the asymptotic average performance.

In summary, this thesis contributes to the systematic consideration of disturbances within the economic MPC framework going beyond the robust tracking of a nominally optimal behavior. The disturbances are considered explicitly which allows to take advantage of the disturbances from an economic point of view in the MPC problem formulation. We presented novel approaches for robust economic MPC and even took a first step towards stochastic economic MPC. Moreover, we provided a thorough theoretical analysis of the presented approaches. We were able to recover results from nominal economic model predictive control and, beyond that, showed and introduced their respective robust counterparts. This provides not only the possibility to employ conceptual ideas from robust stabilizing MPC within economic MPC, but also a theoretically sound framework for *robust economic model predictive control.*

7.2 Future work

In this thesis, we presented several novel approaches to explicitly consider disturbances in the framework of economic MPC. However, there are still various interesting open questions for which this thesis can be seen as a starting point.

For all of the presented approaches, the performance results derived were based on a particular condition on the terminal cost. Even though we presented constructive methods to derive appropriate quadratic terminal cost functions, these approximations deteriorate the provable performance bound and, thus, it could be of interest to remove these terminal cost functions by adapting ideas established within the field of MPC without terminal conditions, see, e.g., (Grüne, 2013; Grüne and Stieler, 2014; Lorenzen et al., 2017b; Reble and Allgöwer, 2012). In order to prove optimality (and stability) for these approaches, the relation of dissipativity to turnpikes is investigated in (Damm et al., 2014; Faulwasser and Bonvin, 2015, 2017; Grüne and Müller, 2016) and it might be worthwhile to investigate robust counterparts of the turnpike property. As discussed in the introduction, steady-state operation need not be the optimal operating regime in economic MPC. In fact, investigating optimal periodic operation is a growing field of interest (Grüne and Zanon, 2014; Müller and Grüne, 2016; Müller et al., 2015b; Zanon et al., 2016b). A first approach towards MPC without terminal conditions as well as periodic optimality in the robust economic framework is presented in Wabersich (2017), which opens the field for further investigations to follow.

For all the performance results presented in this thesis, asymptotic average performance bounds are proven. However, these results are asymptotic statements and it might be of interest to derive similar bounds for any finite horizon. This is usually referred to as transient average performance. A first concept to derive bounds on transient average performance is investigated in the framework of economic MPC without terminal conditions (Grüne, 2013; Grüne and Stieler, 2014) and could also be of interest in the robust setup. A transient counterpart of the asymptotic average constraints investigated in Section 4.4 is presented for the nominal setup in Müller et al. (2014a). One could reformulate these transient average constraints—similar to the idea in Section 4.4 based on robust optimization—in order to achieve robust satisfaction. Considering both types of average constraints in the setup of Chapter 5 could also be of interest, but would require further modifications in order to account for the growing tubes along the prediction.

For all approaches presented in this thesis, exact measurement of the current state was assumed. However, for most practical applications, this is an unrealistic or even impossible assumption. Thus, robust output MPC seems to by a natural extension to the proposed results (see, e.g., Cannon et al. (2012); Chisci and Zappa (2002); Goulart and Kerrigan (2007); Mayne et al. (2006)). For the case including stochastic information in Chapter 5, this is of particular interest due to possible correlations of current estimates to past measurements (see also Remark 5.1). When considering real-world applications, it could also be interesting to extend the approaches presented in this thesis to take discrete actuators into account (see, e.g., Aguilera and Quevedo (2013); Quevedo et al. (2004); Rawlings and Risbeck (2017)).

Considering the stochastic setup presented in Chapter 5, there are several interesting open questions arising in this context. First, the results were presented for linear systems. In the literature, several concepts of nonlinear stochastic (stabilizing) MPC are provided, see, e.g., (Fagiano and Khammash, 2012; Mesbah, 2016; Mesbah et al., 2014) and the references therein. It may be profitable to extend these approaches to the economic framework. However, one might need to modify the MPC algorithm significantly in order to use nonlinear systems depending on the possibilities to compute the distribution of the error along the prediction horizon in a "growing" fashion as provided in Section 5.1. Starting with the stochastic approach presented in Section 5.4, where the converged distribution on the RCI set is considered, could be a promising starting point. A first step into this direction is presented in Sopasakis et al. (2017), where nonlinear systems are allowed but the disturbance is limited to a finite number of values. Second, computational aspects should be further investigated with respect to the distribution evolution—especially for nonlinear systems—in order to allow for implementation (compare Remark 5.7 on page 74). Third, convergence and stability in robust economic MPC with stochastic information is an interesting question as well. A good starting point for the respective analysis might be provided by the results in (Chatterjee and Lygeros, 2015; Kouvaritakis and Cannon, 2015). It might also be of interest if stability in this setup is related to (some) strict stochastic dissipativity notion and whether this is a reasonable assumption in the sense that interesting problem classes satisfy the respective condition.

The tube-based approaches in Chapter 3 and 4 strongly rely on a good approximation of the invariant error set and on a reasonable choice for the disturbance set. As the set of disturbances is often unknown for many many real-world applications, it might be worthwhile to use ideas from parameter estimation to find a good, i.e., small, approximation of the real disturbance set (see, e.g., Di Cairano (2016); Guay et al. (2015); Lorenzen et al. (2017a); Tanaskovic et al. (2014)).

In conclusion, we see that the results at the intersection of economic MPC and robust MPC presented in this thesis provide various research questions expanding both fields into the novel direction of robust economic model predictive control.

Appendix A
Technical proofs

A.1 Proof of Theorem 3.3

In order to show recursive feasibility of Problem 3.1 in Algorithm 3.2, we make use of the candidate input and disturbance sequences

$$\bar{c}(t+1) = \left\{ c^*(1|t), \ldots, c^*(N-1|t), c_s^c \right\}$$

and

$$\tilde{w}(t+1) = \left\{ w^*(1|t), \ldots, w^*(N-1|t), \hat{w} \right\},$$

with arbitrary $\hat{w} \in \mathbb{W}$. At time $t+1$, $x(t+1) = f_\pi(x(t), c^*(0|t), w(t))$. Note that in general $w(t) \neq w^*(0|t)$. Due to the constraint tightening in (3.9f) as well as by the robust invariance of the terminal region $\mathbb{X}_f$ under c_s^c given in Assumption 3.1(ii), $\bar{c}(t+1)$ and $\tilde{w}(t+1)$ are admissible input and disturbance sequences, respectively, for the optimization problem at time $t+1$. Satisfaction of the constraint $\left(x(t), \pi(x(t), c(t)) \right) \in \mathbb{Z}$, for all $t \in \mathbb{I}_{\geq 0}$, follows from feasibility of the initial state x_0 and constraint (3.9f) for $t = 0$.

Considering the performance bound, we can make use of ideas presented in Raimondo et al. (2009). For arbitrary feasible input and disturbance sequences $\boldsymbol{c}(t)$ and $\boldsymbol{w}(t)$, respectively, it holds that

$$
\begin{aligned}
J_{N+1}^c &\left(x(t), \{\boldsymbol{c}(t), c_s^c\}, \{\boldsymbol{w}(t), \hat{w}\} \right) \\
&= \sum_{k=0}^{N-1} \ell_\pi\left(x(k|t), c(k|t) \right) + \ell_\pi\left(x(N|t), c_s^c \right) + V_f^c\left(f_\pi(x(N|t), c_s^c, \hat{w}) \right) \\
&\leq \sum_{k=0}^{N-1} \ell_\pi\left(x(k|t), c(k|t) \right) + V_f^c(x(N|t)) + \rho\left(|\hat{w}| \right) + \ell_\pi(x_s^c, c_s^c) \\
&= J_N^c\left(x(t), \boldsymbol{c}(t), \boldsymbol{w}(t) \right) + \rho\left(|\hat{w}| \right) + \ell_\pi(x_s^c, c_s^c).
\end{aligned}
\tag{A.1}
$$

The inequality follows by Assumption 3.1(iii) and noting that $x(N|t) \in \mathbb{X}_f$ due to (3.9g). With this intermediate result and noting that $|\hat{w}| \leq w_{\max}$ for all $\hat{w} \in \mathbb{W}$, we can see that

$$
\begin{aligned}
V_{N+1}^c\left(x(t) \right) &\leq \max_{\{\boldsymbol{w}(t), \hat{w}\} \in \mathbb{W}^{N+1}} J_{N+1}^c\left(x(t), \{\boldsymbol{c}^*(t), c_s^c\}, \{\boldsymbol{w}(t), \hat{w}\} \right) \\
&\leq \max_{\boldsymbol{w}(t) \in \mathbb{W}^N} J_N^c\left(x(t), \boldsymbol{c}^*(t), \boldsymbol{w}(t) \right) + \rho(w_{\max}) + \ell_\pi(x_s^c, c_s^c) \\
&= V_N^c\left(x(t) \right) + \rho(w_{\max}) + \ell_\pi(x_s^c, c_s^c),
\end{aligned}
\tag{A.2}
$$

recalling that $\boldsymbol{c}^*(t) := \arg\min_{\boldsymbol{c}(t)} \max_{\boldsymbol{w}(t)} J_N^c(x(t), \boldsymbol{c}(t), \boldsymbol{w}(t))$ subject to (3.9). In Lazar et al. (2008, Theorem 4.3), it is shown that

$$-V_N^c\Big(x(t)\Big) \leq -\ell_\pi\Big(x(t), c^*(0|t)\Big) - V_{N-1}^c\Big(x(t+1)\Big). \tag{A.3}$$

Thus, it follows

$$
\begin{aligned}
V_N^c\Big(x(t+1)\Big) - V_N^c\Big(x(t)\Big) &\leq V_{N-1}^c\Big(x(t+1)\Big) - V_N^c\Big(x(t)\Big) + \rho(w_{\max}) + \ell_\pi(x_s^c, c_s^c) \\
&\leq -\ell_\pi\Big(x(t), c^*(0|t)\Big) + \rho(w_{\sup}) + \ell_\pi(x_s^c, c_s^c),
\end{aligned}
\tag{A.4}
$$

using (A.2) to prove the first and (A.3) to prove the second inequality. We take the average ($\frac{1}{T}\sum_{t=0}^{T-1}$) on both sides of (A.4). Examining the limit inferior as $T \to \infty$, the left hand side vanishes since the terms are finite due to compactness of $\mathbb{Z}$ and continuity of ℓ_π and V_f^c (see Angeli et al. (2012, Theorem 1)). With this, the stated asymptotic average performance bound in (3.13) follows. $\qquad\square$

A.2 Proof of Proposition 3.7

It is sufficient to show that all $(\tilde{x}, \tilde{c}) \in \overline{\mathbb{Z}}_s$ are also in $\overline{\mathcal{Z}}_s$. Obviously, $(\tilde{x}, \tilde{c})$ is a steady-state of the nominal dynamics (3.14). It remains to show that $(\bar{x}(t), \pi(\bar{x}(t), \tilde{c})) \in \mathbb{Z}$ (with $\bar{x}(t)$ as in (3.12)) for all $w(t) \in \mathbb{W}$ and for all $t \in \mathbb{I}_{\geq 0}$. Therefore, we consider the error $\bar{e}(t+1) := f_\pi(\bar{x}(t), \tilde{c}, w(t)) - \tilde{x}$, $e(0) = \bar{x}(0) - \tilde{x}$. The proof follows by induction:

Basis: Set $\bar{x}(0) = \tilde{x}$. Since $(\tilde{x}, \tilde{c}) \in \overline{\mathbb{Z}}$, it holds that $(\bar{x}(0), \pi(\bar{x}(0), \tilde{c})) = (\tilde{x}, \pi(\tilde{x}, \tilde{c})) \in \mathbb{Z}$ (by (3.16)). By definition of the RCI set, it follows that $\bar{e}(1) = f_\pi(\tilde{x}, \tilde{c}, w(0)) - \tilde{x} \in \Omega$ for all $w(0) \in \mathbb{W}$, and accordingly $(\bar{x}(1), \pi(\bar{x}(1), \tilde{c})) = (\tilde{x} + \bar{e}(1), \pi(\tilde{x} + \bar{e}(1), \tilde{c})) \in \mathbb{Z}$.

Inductive step: Assume that $\bar{e}(t) \in \Omega$ and $(\bar{x}(t), \pi(\bar{x}(t), \tilde{c})) \in \mathbb{Z}$. Thus, it holds that $\bar{e}(t+1) = f_\pi(\bar{x}(t), \tilde{c}, w(t)) - \tilde{x} \in \Omega$ for all $w(t) \in \mathbb{W}$, and by (3.16) $(\bar{x}(t+1), \pi(\bar{x}(t+1), \tilde{c})) = (\tilde{x} + \bar{e}(t+1), \pi(\tilde{x} + \bar{e}(t+1), \tilde{c})) \in \mathbb{Z}$, such that the inductive step follows, and we have shown that all $(\tilde{x}, \tilde{c}) \in \overline{\mathbb{Z}}_s$ are also in $\overline{\mathcal{Z}}_s$. $\qquad\square$

A.3 Proof of Theorem 3.9

Recursive feasibility and satisfaction of the pointwise-in-time constraints (2.2) follow from tightening the constraints according to (3.16) as provided in (Bayer et al., 2013; Mayne et al., 2005).

Concerning the average performance statement, the proof follows along the lines of Amrit et al. (2011, Theorem 18). Using the feasible (but in general suboptimal) candidate input sequence $\bar{c}(t+1) = \{c^*(1|t), \ldots, c^*(N-1|t), \bar{\kappa}_f(z(N|t))\}$ as well as the candidate nominal initial state $\bar{z}(0|t+1) = z^*(1|t)$, it holds that

$$
\begin{aligned}
V_N^{\max}(x(t+1)) - V_N^{\max}(x(t)) \leq{} & \sum_{k=0}^{N-2} \max_{\omega\in\Omega} \ell_\pi\Big(z^*(k+1|t) + \omega, c^*(k+1|t)\Big) \\
& + \max_{\omega\in\Omega} \ell_\pi\Big(z^*(N|t) + \omega, \bar{\kappa}_f(z^*(N|t))\Big) \\
& + \overline{V}_f\Big(f_\pi(z^*(N|t), \bar{\kappa}_f(z^*(N|t)), 0)\Big) \\
& - \sum_{k=0}^{N-1} \max_{\omega\in\Omega} \ell_\pi\Big(z^*(k|t) + \omega, c^*(k|t)\Big) - \overline{V}_f\Big(z^*(N|t)\Big).
\end{aligned}
\tag{A.5}
$$

Using Assumption 3.2(iii), it follows that

$$V_N^{\max}(x(t+1)) - V_N^{\max}(x(t)) \leq -\max_{\omega \in \Omega} \ell_\pi\Big(z^*(0|t)+\omega, c^*(0|t)\Big) + \max_{\omega \in \Omega} \ell_\pi(z_s^{\max}+\omega, c_s^{\max}). \quad \text{(A.6)}$$

As in the proof of Theorem 3.3, we take the average on both sides of (A.6). Examining the limit inferior as $T \to \infty$, the left hand side vanishes since the terms are finite due to compactness of $\overline{\mathbb{Z}}$ and continuity of ℓ, π, and $\overline{V}_{\mathrm{f}}$ (see Angeli et al. (2012, Theorem 1)). This yields

$$\limsup_{T \to \infty} \frac{1}{T} \sum_{t=0}^{T-1} \max_{\omega \in \Omega} \ell_\pi\Big(z^*(0|t) + \omega, c^*(0|t)\Big) \leq \max_{\omega \in \Omega} \ell_\pi\Big(z_s^{\max} + \omega, c_s^{\max}\Big). \quad \text{(A.7)}$$

From the definition of the RCI set and the constraints on the nominal initial state in (3.17b), $x(t) \in \{z^*(0|t)\} \oplus \Omega$. With this, we can see that

$$\ell_\pi\Big(x(t), c^*(0|t)\Big) \leq \max_{\omega \in \Omega} \ell_\pi\Big(z^*(0|t) + \omega, c^*(0|t)\Big) \quad \text{(A.8)}$$

for all $t \in \mathbb{I}_{\geq 0}$. Combining (A.7) and (A.8), the performance statement in (3.20) follows. $\square$

A.4 Proof of Theorem 4.7

The proof follows along the lines of the nominal statement in Angeli et al. (2012, Proposition 6.4):

$$
\begin{aligned}
0 &\leq \liminf_{T \to \infty} \frac{1}{T} \Big(\lambda\big(z(0|T)\big) - \lambda\big(z(0|0)\big)\Big) \\
&= \liminf_{T \to \infty} \frac{1}{T} \sum_{t=0}^{T-1} \Big(\lambda\big(z(0|t+1)\big) - \lambda\big(z(0|t)\big)\Big) \\
&\leq \liminf_{T \to \infty} \frac{1}{T} \sum_{t=0}^{T-1} s\big(z(0|t), v(0|t)\big) \\
&\leq -\ell^{\mathrm{int}}(z_s^{\mathrm{int}}, v_s^{\mathrm{int}}) + \liminf_{T \to \infty} \frac{1}{T} \sum_{t=0}^{T-1} \ell^{\mathrm{int}}\big(z(0|t), v(0|t)\big).
\end{aligned}
\quad \text{(A.9)}
$$

Non-negativity of the storage function provides the first inequality, the second inequality follows by (4.10) noting that $z(0|t+1) \in \overline{\mathbb{X}}_{\mathrm{init}}(z(1|t))$. The third inequality follows from Ω-robust dissipativity and $\ell^{\mathrm{int}}(z_s^{\mathrm{int}}, v_s^{\mathrm{int}})$ being constant. $\square$

A.5 Proof of Lemma 4.10

In order to show convexity, we have to prove that

$$\ell^{\mathrm{int}}\Big(\gamma z + (1-\gamma)\bar{z}, \gamma v + (1-\gamma)\bar{v}\Big) \leq \gamma\, \ell^{\mathrm{int}}(z, v) + (1-\gamma)\, \ell^{\mathrm{int}}(\bar{z}, \bar{v}), \quad \text{(A.10)}$$

for all (z, v) and $(\bar{z}, \bar{v}) \in \overline{\mathbb{Z}}$ with $\gamma \in [0, 1]$. We note that $\overline{\mathbb{Z}}$ is also convex (see Kolmanovsky and Gilbert (1998, Theorem 2.1)). It follows for the left hand side in (A.10)

$$\ell^{\mathrm{int}}\Big(\gamma z + (1-\gamma)\bar{z}, \gamma v + (1-\gamma)\bar{v}\Big) = \int_{\omega \in \Omega} \ell\Big(\gamma z + (1-\gamma)\bar{z} + \omega, \gamma v + (1-\gamma)\bar{v} + K\omega\Big)\mathrm{d}\omega. \quad \text{(A.11)}$$

Due to convexity of ℓ and because the integration is a linear operator, it follows, by using $\omega = \gamma\omega + (1-\gamma)\omega$,

$$
\begin{aligned}
\int_{\omega\in\Omega} & \ell\big(\gamma z + (1-\gamma)\bar{z} + \omega, \gamma v + (1-\gamma)\bar{v} + K\omega\big)\mathrm{d}\omega \\
& \leq \gamma \int_{\omega\in\Omega} \ell\big(z+\omega, v+K\omega\big)\mathrm{d}\omega + (1-\gamma)\int_{\omega\in\Omega}\ell\big(\bar{z}+\omega, \bar{v}+K\omega\big)\mathrm{d}\omega,
\end{aligned}
\tag{A.12}
$$

leading to (A.10) and providing convexity. $\qquad\square$

A.6 Proof of Theorem 4.17

The proof consists of two part: First, we show asymptotic convergence of $x(t)$ to the set $\{z_s^{\text{int}}\} \oplus \Omega$. Second, we provide stability in a neighborhood of this set. The first part of the proof is based on ideas presented in Amrit et al. (2011) and Rawlings and Mayne (2009, Section 3.4.4). Similar to Amrit et al. (2011), we introduce a rotated stage cost and a rotated terminal cost given by

$$
L^{\text{int}}(z,v) := \ell^{\text{int}}(z,v) + \lambda(z) - \lambda\big(f(z,v,0)\big) - \ell^{\text{int}}(z_s^{\text{int}}, v_s^{\text{int}})
\tag{A.13}
$$

and

$$
\overline{V}_{\mathrm{f}}^{\text{int}}(z) := V_{\mathrm{f}}^{\text{int}}(z) + \lambda(z) - \lambda(z_s^{\text{int}}) - V_{\mathrm{f}}^{\text{int}}(z_s^{\text{int}}),
\tag{A.14}
$$

respectively, as well as the rotated objective by

$$
\overline{J}_N^{\text{int}}(z,\boldsymbol{v}) := \sum_{k=0}^{N-1} L^{\text{int}}\big(z(k), v(k)\big) + \overline{V}_{\mathrm{f}}^{\text{int}}\big(z(N)\big).
\tag{A.15}
$$

Investigating L^{int}, it holds by means of strict Ω-robust dissipativity that

$$
L^{\text{int}}(z,v) \geq \ell^{\text{int}}(z,v) + \lambda(z) - \max_{x\in\overline{\mathbb{X}}_{\text{init}}(f(z,v,0))}\lambda(x) - \ell^{\text{int}}(z_s^{\text{int}}, v_s^{\text{int}}) = \rho(z - z_s^{\text{int}})
\tag{A.16}
$$

for all $(z,v) \in \overline{\mathbb{Z}}$, since $\max_{x\in\overline{\mathbb{X}}_{\text{init}}(f(z,v,0))}\lambda(x) \geq \lambda(f(z,v,0))$, and in addition, it holds that $L^{\text{int}}(z_s^{\text{int}}, v_s^{\text{int}}) = 0$. By Amrit et al. (2011, Lemma 10) it follows $\rho(z - z_s^{\text{int}}) \geq \gamma(|z - z_s^{\text{int}}|) \geq 0$, with a class $\mathcal{K}$ function γ. Moreover, for the rotated terminal cost holds $\overline{V}_{\mathrm{f}}^{\text{int}}(f(z, \bar{\kappa}_{\mathrm{f}}(z), 0)) - \overline{V}_{\mathrm{f}}^{\text{int}}(z) \leq -L^{\text{int}}(z, \bar{\kappa}_{\mathrm{f}}(z))$. For a given $x(t)$, we denote the rotated objective applying the optimal input sequence $\boldsymbol{v}^*(t)$ and the optimal nominal initial state $z^*(0|t)$ (determined with respect to Problem 4.1, i.e., for the non-rotated objective) by $\overline{V}_N^{\text{int}}(x(t)) = \overline{V}_N^o(z^*(0|t)) = \overline{J}_N^{\text{int}}(z^*(0|t), \boldsymbol{v}^*(t))$.

When examining the rotated objective, it follows that

$$
\begin{aligned}
\overline{J}_N^{\text{int}}(z,\boldsymbol{v}) &= \sum_{k=0}^{N-1} L^{\text{int}}\big(z(k), v(k)\big) + \overline{V}_{\mathrm{f}}^{\text{int}}\big(z(N)\big) \\
&= \sum_{k=0}^{N-1} \ell^{\text{int}}\big(z(k), v(k)\big) + V_{\mathrm{f}}^{\text{int}}\big(z(N)\big) \\
&\quad - N\ell^{\text{int}}(z_s^{\text{int}}, v_s^{\text{int}}) - V_{\mathrm{f}}(z_s^{\text{int}}) - \lambda(z_s^{\text{int}}) + \lambda\big(z(0)\big) \\
&= J_N^{\text{int}}(z,\boldsymbol{v}) - N\ell^{\text{int}}(z_s^{\text{int}}, v_s^{\text{int}}) - V_{\mathrm{f}}(z_s^{\text{int}}) - \lambda(z_s^{\text{int}}) + \lambda\big(z(0)\big),
\end{aligned}
\tag{A.17}
$$

where the second equality follows by a telescoping series with respect to the storage function λ. In the nominal case, the nominal initial state $z(0)$ is not an optimization variable but fixed. Thus, the objective J_N^{int} and the rotated objective $\overline{J}_N^{\text{int}}$ only differ by a constant term and yield the same minimizer taking the constraints in (4.4) into consideration. In our case, the nominal initial state is an optimization variable and, therefore, the optimization problems do not result in the same minimizer, and we cannot directly apply the approach from Amrit et al. (2011). However, using the same steps as in Amrit et al. (2011, Theorem 15) (see also Rawlings and Mayne (2009, Proposition 2.17)), one can show that

$$\alpha_1\Big(\big|z - z_s^{\text{int}}\big|\Big) \leq \overline{V}_N^o(z) \leq \alpha_2\Big(\big|z - z_s^{\text{int}}\big|\Big), \tag{A.18}$$

for all $z \in \overline{\mathbb{X}}_N$, where α_1 and α_2 are class $\mathcal{K}$ functions.

Following the standard proof, we introduce the candidate input sequence $\tilde{\boldsymbol{v}}(t+1) = \{v^*(1|t), \ldots, v^*(N-1|t), \bar{\kappa}_f(z^*(N|t))\}$ and the nominal initial state candidate $\tilde{z}(0|t+1) = z^*(1|t) = f(z^*(0|t), v^*(0|t), 0)$. With those, we can derive

$$
\begin{aligned}
\overline{V}_N^o(z^*(0|t+1)) &= J_N^{\text{int}}\Big(z^*(0|t+1), \boldsymbol{v}^*(t+1)\Big) \\
&\quad + \lambda\Big(z^*(0|t+1)\Big) - N\ell^{\text{int}}(z_s^{\text{int}}, v_s^{\text{int}}) - V_f(z_s^{\text{int}}) - \lambda(z_s^{\text{int}}) \\
&\leq J_N^{\text{int}}\Big(\tilde{z}(0|t+1), \tilde{\boldsymbol{v}}(t+1)\Big) \\
&\quad + \lambda\Big(z^*(0|t+1)\Big) - N\ell^{\text{int}}(z_s^{\text{int}}, v_s^{\text{int}}) - V_f(z_s^{\text{int}}) - \lambda(z_s^{\text{int}}) \\
&\leq J_N^{\text{int}}\Big(\tilde{z}(0|t+1), \tilde{\boldsymbol{v}}(t+1)\Big) \\
&\quad + \max_{x \in \overline{\mathbb{X}}_{\text{init}}(\tilde{z}(0|t+1))} \lambda(x) - N\ell^{\text{int}}(z_s^{\text{int}}, v_s^{\text{int}}) - V_f(z_s^{\text{int}}) - \lambda(z_s^{\text{int}})
\end{aligned}
\tag{A.19}
$$

The first inequality follows by noting that the candidate solution is in general a suboptimal solution for the objective (4.5); the second inequality follows since $z^*(0|t+1) \in \overline{\mathbb{X}}_{\text{init}}(\tilde{z}(0|t+1))$ and $\tilde{z}(0|t+1) = z^*(1|t)$. With this bound, we can see that

$$
\begin{aligned}
\overline{V}_N^o\Big(z^*(0|t+1)\Big) - \overline{V}_N^o\Big(z^*(0|t)\Big) &\leq J_N^{\text{int}}\Big(\tilde{z}(0|t+1), \tilde{\boldsymbol{v}}(t+1)\Big) - J_N^{\text{int}}\Big(z^*(0|t), \boldsymbol{v}^*(t)\Big) \\
&\quad + \max_{x \in \overline{\mathbb{X}}_{\text{init}}(\tilde{z}(0|t+1))} \lambda(x) - \lambda(z^*(0|t)) \\
&= \ell^{\text{int}}\Big(z^*(N|t), \bar{\kappa}_f(z^*(N|t))\Big) - \ell^{\text{int}}\Big(z^*(0|t), v^*(0|t)\Big) \\
&\quad + V_f^{\text{int}}\Big(f\big(z^*(N|t), \bar{\kappa}_f(z^*(N|t)), 0\big)\Big) - V_f^{\text{int}}\Big(z^*(N|t)\Big) \\
&\quad + \max_{x \in \overline{\mathbb{X}}_{\text{init}}(\tilde{z}(0|t+1))} \lambda(x) - \lambda\Big(z^*(0|t)\Big) \\
&\leq -\ell^{\text{int}}\Big(z^*(0|t), v^*(0|t)\Big) + \ell^{\text{int}}(z_s^{\text{int}}, v_s^{\text{int}}) \\
&\quad + \max_{x \in \overline{\mathbb{X}}_{\text{init}}(\tilde{z}(0|t+1))} \lambda(x) - \lambda\Big(z^*(0|t)\Big) \\
&\leq -\gamma\Big(\big|z^*(0|t) - z_s^{\text{int}}\big|\Big).
\end{aligned}
\tag{A.20}
$$

The first inequality follows by (A.19), the second by Assumption 4.1(iii), and the third by strict Ω-robust dissipativity.

As provided in Rawlings and Mayne (2009, Section 3.4.4), $\overline{V}_N^{\text{int}}(x) = 0$ for all $x \in \{z_s^{\text{int}}\}\oplus\Omega$. By means of properties (A.18) and (A.20), we can find a class $\mathcal{KL}$ function β such that $|z^*(0|t) - z_s^{\text{int}}| \leq \beta(|z^*(0|0) - z_s^{\text{int}}|, t)$ (see, e.g., Jiang and Wang (2001); Rawlings and Mayne (2012)). For all $t \in \mathbb{I}_{\geq 0}$, $x(t) = z^*(0|t) + e(t)$, where $e(t) \in \Omega$, so that $|x(t)|_{\{z_s^{\text{int}}\}\oplus\Omega} = |z^*(0|t) + e(t)|_{\{z_s^{\text{int}}\}\oplus\Omega} \leq |z^*(0|t) - z_s^{\text{int}}|$. Hence,

$$\left|x(t)\right|_{\{z_s^{\text{int}}\}\oplus\Omega} \leq \left|z^*(0|t) - z_s^{\text{int}}\right| \leq \beta\left(\left|z^*(0|0) - z_s^{\text{int}}\right|, t\right). \tag{A.21}$$

Thus, we have asymptotic convergence of $x(t)$ to the set $\{z_s^{\text{int}}\} \oplus \Omega$.

We note that the above relation does *not* yet provide asymptotic stability (see discussion in Rawlings and Mayne (2009, Section 3.4.4)), but we must establish stability in a neighborhood of $\{z_s^{\text{int}}\} \oplus \Omega$. Therefore, consider

$$z_{\min} := z_s^{\text{int}} + x(0) - y^*, \tag{A.22}$$

where $x(0)$ is given and

$$y^* := \underset{y \in \{z_s^{\text{int}}\}\oplus\Omega}{\arg\min} \left|y - x(0)\right|. \tag{A.23}$$

With this relation as depicted in Figure A.1, it follows that $z_{\min} \in \{x(0)\} \oplus (-\Omega)$ since $x(0) - z_{\min} = y^* - z_s^{\text{int}} \in \Omega$. Thus, using the geometric connection, we can show that

$$\left|x(0)\right|_{\{z_s^{\text{int}}\}\oplus\Omega} = \left|z_{\min} - z_s^{\text{int}}\right| = \left|x(0) - y^*\right|. \tag{A.24}$$

In general, the state $z_{\min}$ will not be the optimal nominal initial state, but we can use it in order to show stability.

Now, we consider all initial states $x(0) \in (\{z_s^{\text{int}}\} \oplus \Omega \oplus \overline{\mathcal{B}}_\epsilon) \cap \mathbb{X}$, that is all states in an ϵ-neighborhood of the set to be stabilized. All possible nominal initial states $z(0|0)$ are limited to the set $(\{x(0)\} \oplus (-\Omega)) \cap \overline{\mathbb{X}}$ and, thus, to the set $(\overline{\mathbb{X}}_{\text{init}}(z_s^{\text{int}}) \oplus \overline{\mathcal{B}}_\epsilon) \cap \overline{\mathbb{X}}$. Taking (A.24) into account and recalling that $x(0) \in (\{z_s^{\text{int}}\} \oplus \Omega \oplus \overline{\mathcal{B}}_\epsilon) \cap \mathbb{X}$, it holds that

$$\left|z_{\min} - z_s^{\text{int}}\right| = \left|x(0)\right|_{\{z_s^{\text{int}}\}\oplus\Omega} \leq \epsilon. \tag{A.25}$$

With respect to the bound on ϵ stated, $z_{\min} \in \overline{\mathbb{X}}_{\text{f}} \subseteq \overline{\mathbb{X}}$, and thus, we can derive a feasible state and input sequence by means of repeated application of the terminal auxiliary controller $\bar{\kappa}_{\text{f}}$, that is, $\mathbf{z}_{\text{f}}(z_{\min}, \bar{\kappa}_{\text{f}})$ and $\mathbf{v}_{\text{f}}(z_{\min}, \bar{\kappa}_{\text{f}})$. Using Amrit et al. (2011, Lemma 11), it follows that

$$\overline{V}_{\text{f}}^{\text{int}}(z_{\min}) \geq \sum_{k=0}^{M-1} L^{\text{int}}\left(z_{\text{f}}(k; z_{\min}, \bar{\kappa}_{\text{f}}), v_{\text{f}}(k; z_{\min}, \bar{\kappa}_{\text{f}})\right) + \overline{V}_{\text{f}}^{\text{int}}\left(z_{\text{f}}(M; z_{\min}, \bar{\kappa}_{\text{f}})\right), \tag{A.26}$$

that is, the rotated terminal cost is an upper bound for the rotated cost under the terminal controller. Moreover, one can show that the rotated terminal cost is positive definite on $\overline{\mathbb{X}}_{\text{f}}$ with respect to $z = z_s^{\text{int}}$, and by Amrit et al. (2011, Lemma 13), there exists a class $\mathcal{K}$ function $\hat{\gamma}$ such that $\overline{V}_{\text{f}}^{\text{int}}(z) \leq \hat{\gamma}(|z - z_s^{\text{int}}|)$ for all $z \in \overline{\mathbb{X}}_{\text{f}}$.

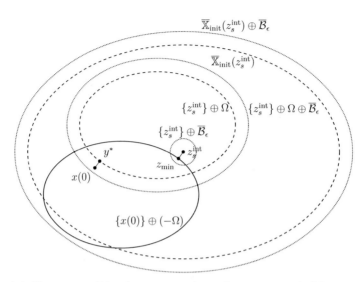

Figure A.1: Illustration of the chosen nominal initial state $z_{\min}$ and $x(0)$ as well as the relation to the other sets considered within the proof.

With this, we can derive for all $x(0) \in (\{z_s^{\text{int}}\} \oplus \Omega \oplus \overline{B}_\epsilon) \cap \mathbb{X}$

$$
\begin{aligned}
V_N^{\text{int}}\big(x(0)\big) = \overline{V}_N^o\big(z^*(0|0)\big) &\leq \overline{V}_N^o\big(z^*(0|0)\big) + \lambda(z_s^{\text{int}}) - \lambda\big(z^*(0|0)\big) \\
&= J_N^{\text{int}}\big(z^*(0|0), \boldsymbol{v}^*(t)\big) - N\ell^{\text{int}}(z_s^{\text{int}}, v_s^{\text{int}}) - V_{\text{f}}^{\text{int}}(z_s^{\text{int}}) \\
&\leq J_N^{\text{int}}\big(z_{\min}, \boldsymbol{v}_{\text{f}}(z_{\min}, \bar{\kappa}_{\text{f}})\big) - N\ell^{\text{int}}(z_s^{\text{int}}, v_s^{\text{int}}) - V_{\text{f}}^{\text{int}}(z_s^{\text{int}}) \\
&= \overline{J}_N^{\text{int}}\big(z_{\min}, \boldsymbol{v}_{\text{f}}(z_{\min}, \bar{\kappa}_{\text{f}})\big) - \lambda(z_{\min}) + \lambda(z_s^{\text{int}}) \\
&\leq \overline{V}_{\text{f}}^{\text{int}}(z_{\min}) - \lambda(z_{\min}) + \lambda(z_s^{\text{int}}) \\
&\leq \hat{\gamma}\big(|z_{\min} - z_s^{\text{int}}|\big) + \alpha_\lambda\big(|z_{\min} - z_s^{\text{int}}|\big) \\
&= \overline{\alpha}\big(|z_{\min} - z_s^{\text{int}}|\big),
\end{aligned}
\tag{A.27}
$$

where α_λ and $\overline{\alpha} := \hat{\gamma} + \alpha_\lambda$ are class $\mathcal{K}$ functions. A few comments considering this derivation are in order: The first inequality follows by (4.23), which is satisfied since $z^*(0|0) \in (\overline{\mathbb{X}}_{\text{init}}(z_s^{\text{int}}) \oplus \overline{B}_\epsilon) \cap \mathbb{X}$. The second inequality holds by suboptimality of $z_{\min}$ and by using the terminal controller, which provides a feasible solution since $z_{\min} \in \overline{\mathbb{X}}_{\text{f}}$ (see (A.25)). The third inequality follows by (A.26) with $M = N$. The class $\mathcal{K}$ function bound for $\lambda(z_s^{\text{int}}) - \lambda(z_{\min})$ can be derived by Amrit et al. (2011, Lemma 10). Using (A.18) and the geometric relation (A.24), it follows that for all $z^*(0|0) \in (\overline{\mathbb{X}}_{\text{init}}(z_s^{\text{int}}) \oplus \overline{B}_\epsilon) \cap \overline{\mathbb{X}}$

$$
\Big|z^*(0|0) - z_s^{\text{int}}\Big| \leq \alpha_1^{-1}\Big(\overline{\alpha}\big(|x(0)|_{\{z_s^{\text{int}}\}\oplus\Omega}\big)\Big),
\tag{A.28}
$$

which leads to stability.

Thus, by asymptotic convergence and stability, $\{z_s^{\text{int}}\} \oplus \Omega$ is asymptotically stable for the real system under the additional assumption stated in (4.23). $\qquad\qquad\square$

A.7 Proof of Theorem 4.20

The proof follows along the proof of Theorem 4.17 and the proof of Rawlings and Mayne (2009, Proposition 3.15) (using comparison functions). Since we have a nominal system, we can show—using strict robust dissipativity which is equivalent to strict dissipativity of the nominal system—that the rotated value function can be used as a Lyapunov function for the nominal system. Moreover, the discussion in Remark 4.18 becomes obsolete since the nominal initial state is no longer an optimization variable and the rotated objective has the same minimizer as the original objective (subject to the same constraints). Thus, z_s^{int} is asymptotically stable for the nominal closed-loop system (4.24a) with region of attraction $\overline{\mathbb{X}}_N$. This means there exists a class $\mathcal{KL}$ function β such that

$$\left| z(t) - z_s^{\text{int}} \right| \leq \beta\left(\left| z(0) - z_s^{\text{int}} \right|, t \right), \qquad (A.29)$$

for all $t \in \mathbb{I}_{\geq 0}$.

Regarding stability of the composite system, it follows similarly as in (A.21) that $|x(t)|_{\{z_s^{\text{int}}\}\oplus\Omega} \leq \beta(|z(0) - z_s^{\text{int}}|, t)$. Using this relation, we obtain

$$\begin{aligned}
\left| \left(z(t), x(t) \right) \right|_{\mathcal{A}} &\leq \left| z(t) - z_s^{\text{int}} \right| + \left| x(t) \right|_{\{z_s^{\text{int}}\}\oplus\Omega} \\
&\leq 2\beta\left(\left| z(0) - z_s^{\text{int}} \right|, t \right) \leq 2\beta\left(\left| (z(0), x(0)) \right|_{\mathcal{A}}, t \right)
\end{aligned} \qquad (A.30)$$

for all $t \in \mathbb{I}_{\geq 0}$. This proves that the set $\mathcal{A}$ is asymptotically stable for the closed-loop composite system (4.24) with the region of attraction $\overline{\mathbb{X}}_N \times (\overline{\mathbb{X}}_N \oplus \Omega)$. $\qquad\square$

A.8 Proof of Theorem 4.23

We consider the candidate solution for the nominal open-loop predictions at time $t + 1$, given by $\tilde{z}(t + 1) = \{z^*(1|t), \ldots, z^*(N|t), f(z^*(N|t), \bar{\kappa}_f(z^*(N|t)), 0)\}$ and $\tilde{v}(t + 1) = \{v^*(1|t), \ldots, v^*(N-1|t), \bar{\kappa}_f(z^*(N|t))\}$. Recursive feasibility and satisfaction of the pointwise-in-time constraints follow by the appropriate tightening of $\overline{\mathbb{Z}}$.

Considering recursive feasibility of the average constraints, we take the sequences $\tilde{z}(t+1)$ and $\tilde{v}(t + 1)$ into account for which follows

$$\begin{aligned}
\sum_{k=0}^{N-1} \overline{h}\left(\tilde{z}(k), \tilde{v}(k) \right) &= \sum_{k=0}^{N-1} \overline{h}\left(z^*(k|t), v^*(k|t) \right) + \overline{h}\left(z^*(N|t), \bar{\kappa}_f(z^*(N|t)) \right) - \overline{h}\left(z^*(0|t), v^*(0|t) \right) \\
&\in \overline{\mathbb{Y}}_t \oplus \mathbb{R}^p_{\leq 0} \oplus \left\{ -\overline{h}\left(z^*(0|t), v^*(0|t) \right) \right\} = \overline{\mathbb{Y}}_{t+1}.
\end{aligned} \qquad (A.31)$$

The set relation is based on (4.33) in Assumption 4.3. Thus, the additional constraint (4.34) is recursively feasible with the set $\overline{\mathbb{Y}}_t$ defined in (4.35).

Next, we want to focus on satisfaction of the average constraint (4.27). Therefore, we solve the recursion for $\overline{\mathbb{Y}}_t$, i.e.,

$$\overline{\mathbb{Y}}_t = \overline{\mathbb{Y}}_{00} \oplus (t+1)\mathbb{R}^p_{\leq 0} \oplus \left\{ -\sum_{k=0}^{t-1} \overline{h}\Big(z^*(0|k), v^*(0|k)\Big) \right\} \tag{A.32}$$

Following the proof of Angeli et al. (2012, Theorem 5), we can rewrite (4.34) as

$$\sum_{k=0}^{N-1} \overline{h}\Big(z^*(k|t), v^*(k|t)\Big) + \sum_{k=0}^{t-1} \overline{h}\Big(z^*(0|t), v^*(0|t)\Big) \in \overline{\mathbb{Y}}_{00} \oplus (t+1)\mathbb{R}^p_{\leq 0}. \tag{A.33}$$

We note that the first sum consists of N terms only. Moreover, each of these terms can be bounded due to compactness of $\overline{\mathbb{Z}}$ and continuity of h. Considering any infinite subsequence $t_n \subseteq \mathbb{I}_{\geq 0}$ such that a limit is admitted for $s_{\overline{h}}(t_n) := \frac{1}{t_n}\sum_{k=0}^{t_n-1} \overline{h}(z^*(0|k), v^*(0|k))$, we can see that

$$\lim_{n\to\infty} s_{\overline{h}}(t_n) \in \lim_{n\to\infty} \frac{\left(\overline{\mathbb{Y}}_{00} \oplus (t_n+1)\mathbb{R}^p_{\leq 0}\right)}{t_n} = \mathbb{R}^p_{\leq 0}, \tag{A.34}$$

where the set-limit has to be understood in the sense of Goebel et al. (2012, Definition 5.1).

Up to now, we have merely shown that $\mathrm{Av}[\overline{h}(z, v)] \subseteq \mathbb{R}^p_{\leq 0}$. In order to show satisfaction of the average constraint (4.27) for the real closed-loop system, we have to recall that each element of $\mathrm{Av}[\overline{h}(z, v)]$ is a limit point of the sequence $s_{\overline{h}}(t) = \frac{1}{t}\sum_{k=0}^{t-1} \overline{h}(z^*(0|k), v^*(0|k))$ (when taking the limit along a specific subsequence t_n). Due to the fact that $\mathrm{Av}[\overline{h}(z, v)] \subseteq \mathbb{R}^p_{\leq 0}$, it follows that $\limsup_{t\to\infty}[s_{\overline{h}}(t)]_i \leq 0$, for all $i \in \mathbb{I}_{[1,p]}$.

From the definition of $\overline{h}$ in (4.31) and the input parametrization (2.14), it follows that the closed-loop system (4.36) satisfies $h(x(t), u(t)) \leq \overline{h}(z^*(0|t), v^*(0|t))$ for all $t \in \mathbb{I}_{\geq 0}$. Hence, we have $s_h(t) \leq s_{\overline{h}}(t)$ for all $t \in \mathbb{I}_{\geq 0}$, where $s_h(t) := \frac{1}{t}\sum_{k=0}^{t-1} h(x(k), u(k))$, which in particular implies that $\limsup_{t\to\infty}[s_h(t)]_i \leq \limsup_{t\to\infty}[s_{\overline{h}}(t)]_i \leq 0$, for all $i \in \mathbb{I}_{[1,p]}$. However, this means that $\mathrm{Av}[h(x, u)] \subseteq \mathbb{R}^p_{\leq 0}$, i.e., the closed-loop system satisfies the average constraints as required.

The last statement—the bound on the asymptotic performance—follows directly from the proof of Theorem 3.9. $\qquad\square$

A.9 Proof of Lemma 5.2

Let $v_i^\top$ and λ_i, $i = 1, \ldots, n$ be the generalized normalized left eigenvectors and the generalized eigenvalues of A_{cl}, respectively. We note that the result can be shown for all A_{cl}, but for ease of presentation we restrict the following discussion to the case of all λ_i having equal geometric and algebraic multiplicity. Hence, for each $\alpha \in \mathbb{R}^n$, $\alpha^\top = \sum_{i=1}^n \alpha_i v_i^\top$, where $\alpha_i \in \mathbb{R}$, since the linear span of all (left) eigenvectors of A_{cl} is the whole $\mathbb{R}^n$.

Next, we define $\tilde{e}_{k+1} = \sum_{i=0}^k A_{\mathrm{cl}}^i w_i$, $\tilde{e}_0 = 0$. Here, the order of summation is different to the one that would follow from the error definition in (5.6) which provides $e(k+1|t) = \sum_{i=0}^k A_{\mathrm{cl}}^{k-i} w_i$. Moreover, we introduce the infinite sum $\tilde{e} = \sum_{i=0}^\infty A_{\mathrm{cl}}^i w_i$, which exists since all w_i are bounded and A_{cl} is asymptotically stable.

In our problem, we have an infinite dimensional (countable) product probability space

$$\left(\mathbb{W}^{\mathbb{I}_{\geq 1}}, \mathcal{B}_\mathbb{W}^{\mathbb{I}_{\geq 1}}, \underbrace{\bigotimes_{i\geq 0} \mathbb{P}_w}_{=:\mathbb{P}}\right), \tag{A.35}$$

with the sample space $\mathbb{W}^{\mathbb{I}_{\geq 1}}$, the Borel σ-algebra $\mathcal{B}_{\mathbb{W}}^{\mathbb{I}_{\geq 1}}$, and the probability measure $\mathbb{P}$. This is possible due to the Kolmogorov extension theorem (see, e.g., Klenke (2014, Theorem 14.36)), which states that for each stochastic process (under some very mild assumptions) one can find an appropriate associated probability space.

As an intermediate step, we show *convergence in probability*. Using the introduced α, it holds with with $\varepsilon > 0$

$$
\mathbb{P}\left\{\sup_{k \geq \bar{n}} \left|\alpha^\top \tilde{e}_{k+1} - \alpha^\top \tilde{e}\right| \geq \varepsilon\right\} = \mathbb{P}\left\{\sup_{k \geq \bar{n}} \left|\alpha^\top \sum_{i=k+1}^{\infty} A_{\mathrm{cl}}^i w_i\right| \geq \varepsilon\right\}
$$

$$
= \mathbb{P}\left\{\sup_{k \geq \bar{n}} \left|\sum_{i=k+1}^{\infty} \sum_{j=1}^{n} \alpha_j \lambda_j^i v_j^\top w_i\right| \geq \varepsilon\right\}
$$

$$
\leq \mathbb{P}\left\{\sup_{k \geq \bar{n}} \sum_{i=k+1}^{\infty} (\max_{l=1,\dots,n} |\lambda_l|)^i \underbrace{\left|\sum_{j=1}^{n} \alpha_j v_j^\top w_i\right|}_{\leq c(\alpha)} \geq \varepsilon\right\}.
$$

$$(A.36)$$

The upper bound $c(\alpha)$ exists as α is fixed and $\mathbb{W}$ is bounded. With this bound and by denoting $\max_{l=1,\dots,n} |\lambda_l| = \bar{\lambda}$, we can show that for all $\alpha \in \mathbb{R}^n$

$$
\sup_{k \geq \bar{n}} c(\alpha) \sum_{i=k+1}^{\infty} \bar{\lambda}^i \leq \sup_{k \geq \bar{n}} c(\alpha) \left(\frac{1}{1-\bar{\lambda}} - \frac{1 - \bar{\lambda}^{\bar{n}+1}}{1-\bar{\lambda}}\right)
$$

$$
= c(\alpha) \frac{\bar{\lambda}^{\bar{n}+1}}{1-\bar{\lambda}} \to 0, \quad \text{for } \bar{n} \to \infty.
$$

$$(A.37)$$

This approach of a geometric series holds since all eigenvalues of A_{cl} are strictly inside the unit disc, and hence $\bar{\lambda} < 1$. With the result in (A.37), we can derive for all $\varepsilon > 0$ and for all $\alpha \in \mathbb{R}^n$

$$
\mathbb{P}\left\{\sup_{k \geq \bar{n}} \left|\alpha^\top \tilde{e}_{k+1} - \alpha^\top \tilde{e}\right| \geq \varepsilon\right\} \overset{\bar{n} \to \infty}{\to} 0.
$$

$$(A.38)$$

This provides convergence in probability, i.e., $\alpha^\top \tilde{e}_{\bar{n}+1} \overset{\mathbb{P}}{\to} \alpha^\top \tilde{e}$. Using standard results for stochastic systems (see, e.g., Poznyak (2009)), *convergence in distribution* follows directly from convergence in probability, hence,

$$
\alpha^\top \tilde{e}_{\bar{n}+1} \overset{\mathrm{d}}{\to} \alpha^\top \tilde{e}, \quad \forall \alpha \in \mathbb{R}^n.
$$

$$(A.39)$$

Using the Cramér-Wold Theorem (see, e.g., Klenke (2014, Theorem 15.56)), we can see that

$$
\tilde{e}_{\bar{n}+1} \overset{\mathrm{d}}{\to} \tilde{e}.
$$

$$(A.40)$$

As the disturbances are i.i.d., the distribution for $\tilde{e}_k$ is the same as for $e(k|t)$. Thus, one can also derive that $e(\bar{n}+1|t) = \sum_{k=0}^{\bar{n}} A_{\mathrm{cl}}^{\bar{n}-k} w_k \to \tilde{e}$. Since the stage cost ℓ is continuous and, thus, bounded on $\mathbb{Z}$, it follows (using results from the Portmanteau theorem, see, e.g., Klenke (2014, Theorem 13.16), and noting that we consider an infinite dimensional probability space)

$$
\mathbb{E}\left\{\tilde{\ell}(e_{\bar{n}})\right\} \to \mathbb{E}\left\{\tilde{\ell}(\tilde{e})\right\}, \quad \text{as } \bar{n} \to \infty,
$$

$$(A.41)$$

where $\tilde{\ell}(e) = \ell(z+e, Ke+v)$ with z and v being constant. $\qquad\square$

A.10 Proof of Theorem 5.5

We split the proof into three parts, proving recursive feasibility, constraint satisfaction, as well as the performance bound separately from each other.

Recursive feasibility can be shown along the lines of the setup in Chisci et al. (2001). Even though a different input parametrization is applied, we can still follow the same steps. We use the candidate input given by $\tilde{v}(t+1) = \{v^*(1|t) + Kw(t), \ldots, v^*(N-1|t) + KA_{\text{cl}}^{N-2}w(t), K(z^*(N|t) + A_{\text{cl}}^{N-1}w(t) - z_s^\infty) + v_s^\infty\}$, which is possible since $w(t)$ is known at time $t+1$.

In order to prove recursive feasibility, consider the candidate sequences $\tilde{z}(t+1)$ and $\tilde{v}(t+1)$. One can see that for all $w(t) \in \mathbb{W}$, it holds

$$\left(\tilde{z}(k), \tilde{v}(k)\right) \in \overline{\mathbb{Z}}_{k+1} \oplus \left(A_{\text{cl}}^k \mathbb{W} \times KA_{\text{cl}}^k \mathbb{W}\right) \subseteq \overline{\mathbb{Z}}_k, \ \forall k \in \mathbb{I}_{[0,N-2]}. \tag{A.42}$$

Moreover, we know by (5.17d) that $z^*(N|t) \in \overline{\mathbb{X}}_{\text{f}}$. By definition of the maximal output admissible set O_{max} in (2.29), it follows that $(z^*(N|t) + A_{\text{cl}}^{N-1}w(t), K(z^*(N|t) + A_{\text{cl}}^{N-1}w(t) - z_s^\infty) + v_s^\infty) \in \overline{\mathbb{Z}}_{N-1}$ for all $w(t) \in \mathbb{W}$. Thus, recursive feasibility of the pointwise-in-time constraints (5.17c) is guaranteed for the candidate solution. Considering the terminal constraint (5.17d), the terminal region defined by $O_{\text{max}} \ominus \Omega_N$ can be proven to be robust invariant with respect to $A_{\text{cl}}^N \mathbb{W}$ under application of the terminal controller which proves that $A_{\text{cl}}z^*(N|t) + B(v_s^\infty - Kz_s^\infty) + A_{\text{cl}}^N w(t) \in \overline{\mathbb{X}}_{\text{f}}$ for all $w(t) \in \mathbb{W}$.

Satisfaction of the pointwise-in-time constraints follows directly from the discussion on recursive feasibility since $(z^*(0|t), v^*(0|t)) = (x(t), u(t)) \in \mathbb{Z}$ for all times $t \in \mathbb{I}_{\geq 0}$.

With regard to the *bound on the asymptotic average performance* in (5.20), we introduce the suboptimal value function $\tilde{V}(x(t)) = J_N^{\text{int}}(x(t), \tilde{v}(t))$ resulting from applying the candidate solution $\tilde{v}(t)$. This leads to

$$
\begin{aligned}
V_N^{\text{int}}\left(x(T)\right) - V_N^{\text{int}}\left(x(0)\right) &= \sum_{t=0}^{T-1}\left[V_N^{\text{int}}\left(x(t+1)\right) - V_N^{\text{int}}\left(x(t)\right)\right] \\
&\leq \sum_{t=0}^{T-1}\left[\tilde{V}\left(x(t+1)\right) - V_N^{\text{int}}\left(x(t)\right)\right].
\end{aligned}
\tag{A.43}
$$

Employing the conditional expectation given $x(0)$, it follows with the law of iterated expectation as T is fixed and by monotonicity that

$$\mathbb{E}^0\left\{V_N^{\text{int}}\left(x(T)\right) - V_N^{\text{int}}\left(x(0)\right)\right\} \leq \mathbb{E}^0\left\{\sum_{t=0}^{T-1}\mathbb{E}\left\{\tilde{V}\left(x(t+1)\right)\middle| x(t)\right\} - V_N^{\text{int}}\left(x(t)\right)\right\}, \tag{A.44}$$

using that all terms are bounded due to boundedness of ℓ_k^{int} on $\overline{\mathbb{Z}}_k$ and continuity of $\overline{V}_{\text{f}}^{\text{int}}$ on $\overline{\mathbb{X}}_{\text{f}}$. Considering the right hand side of (A.44), we can compute for all $t \in \mathbb{I}_{[0,T-1]}$

$$
\begin{aligned}
&\mathbb{E}\left\{\tilde{V}\left(x(t+1)\right)\middle| x(t)\right\} - V_N^{\text{int}}\left(x(t)\right) \\
&= \sum_{k=0}^{N-2}\mathbb{E}\left\{\ell_k^{\text{int}}\left(z^*(k+1|t) + A_{\text{cl}}^k w(t), v^*(k+1|t) + KA_{\text{cl}}^k w(t)\right)\middle| x(t)\right\} \\
&\quad + \mathbb{E}\left\{\ell_{N-1}^{\text{int}}\left(z^*(N|t) + A_{\text{cl}}^{N-1}w(t), K(z^*(N|t) + A_{\text{cl}}^{N-1}w(t) - z_s^\infty) + v_s^\infty\right)\middle| x(t)\right\} \\
&\quad + \mathbb{E}\left\{\overline{V}_{\text{f}}^{\text{int}}\left(A_{\text{cl}}z^*(N|t) + B(v_s^\infty - Kz_s^\infty) + A_{\text{cl}}^N w(t)\right)\middle| x(t)\right\} \\
&\quad - \sum_{k=0}^{N-1}\ell_k^{\text{int}}\left(z^*(k|t), v^*(k|t)\right) - \overline{V}_{\text{f}}^{\text{int}}\left(z^*(t)\right).
\end{aligned}
\tag{A.45}
$$

123

With the definition of ℓ_k^{int} in (5.11), it follows that

$$\mathbb{E}\left\{\ell_k^{\text{int}}\left(z^*(k+1|t)+A_{\text{cl}}^k w(t), v^*(k+1|t)+KA_{\text{cl}}^k w(t)\right)\Big|\, x(t)\right\}$$

$$= \int_{\mathbb{W}}\int_{\Omega_k}\ell\left(z^*(k+1|t)+\epsilon+A_{\text{cl}}^k\omega, K(\epsilon+A_{\text{cl}}^k\omega)+v^*(k+1|t)\right)\rho_{\Omega_k}(\epsilon)\rho_{\mathbb{W}}(\omega)\,\mathrm{d}\epsilon\,\mathrm{d}\omega$$

$$= \int_{\Omega_{k+1}}\ell\left(z^*(k+1|t)+\epsilon, K\epsilon+v^*(k+1|t)\right)\rho_{\Omega_{k+1}}(\epsilon)\,\mathrm{d}\epsilon \tag{A.46}$$

$$= \ell_{k+1}^{\text{int}}\left(z^*(k+1|t), v^*(k+1|t)\right).$$

The connection stated above follows by the recursions (2.24) and (5.9) as well as since the disturbances are i.i.d.. With this, we can derive from (A.45) that

$$\mathbb{E}\left\{\tilde{V}\left(x(t+1)\right)\Big|\, x(t)\right\} - V_N^{\text{int}}\left(x(t)\right)$$

$$= \ell_N^{\text{int}}\left(z^*(N|t), K(z^*(N|t)-z_s^\infty)+v_s^\infty\right) - \ell\left(z^*(0|t), v^*(0|t)\right) \tag{A.47}$$

$$+ \mathbb{E}\left\{\overline{V}_{\text{f}}^{\text{int}}\left(A_{\text{cl}}z^*(N|t)+B(v_s^\infty-Kz_s^\infty)+A_{\text{cl}}^N w(t)\right)\Big|\, x(t)\right\} - \overline{V}_{\text{f}}^{\text{int}}\left(z^*(N|t)\right).$$

Thus, using Assumption 5.2, (A.44) and (A.47), it follows that

$$\mathbb{E}^0\left\{V_N^{\text{int}}\left(x(t)\right) - V_N^{\text{int}}\left(x(0)\right)\right\} \le \mathbb{E}^0\left\{\sum_{t=0}^{T-1}\ell_\infty^{\text{int}}\left(z_s^\infty, v_s^\infty\right) - \ell\left(x(t), v^*(0|t)\right)\right\}, \tag{A.48}$$

taking into consideration that in closed loop $x(t) = z^*(0|t)$ (due to (5.17b)). Since $\mathbb{E}^0\{V_N^{\text{int}}(x(0))\} = V_N^{\text{int}}(x(0))$ as well as $\mathbb{E}^0\{\ell_\infty^{\text{int}}(z_s^\infty, v_s^\infty)\} = \ell_\infty^{\text{int}}(z_s^\infty, v_s^\infty)$, we can take the average on both sides of (A.48)

$$\frac{1}{T}\left(\mathbb{E}^0\left\{V_N^{\text{int}}\left(x(T)\right)\right\} - V_N^{\text{int}}\left(x(0)\right)\right) \le \ell_\infty^{\text{int}}\left(z_s^\infty, v_s^\infty\right) - \frac{1}{T}\sum_{t=0}^{T-1}\mathbb{E}^0\left\{\ell\left(x(t), v^*(0|t)\right)\right\}. \tag{A.49}$$

Due to compactness of $\mathbb{Z}(=\overline{\mathbb{Z}}_0)$ and continuity of ℓ and $\overline{V}_{\text{f}}^{\text{int}}$, the terms on the left hand side are finite such that by taking the limit inferior as $T\to\infty$ the left hand side vanishes. Thus, it follows

$$\limsup_{T\to\infty}\frac{1}{T}\sum_{t=0}^{T-1}\mathbb{E}^0\left\{\ell\left(x(t), v^*(0|t)\right)\right\} \le \ell_\infty^{\text{int}}(z_s^\infty, v_s^\infty) \tag{A.50}$$

proving the bound on the asymptotic average performance in (5.20). □

A.11 Proof of Theorem 5.14

Along general lines, we follow the proof of Angeli et al. (2012, Proposition 6.4), however, we have to account for the uncertain disturbance. By local integrability and boundedness of the storage function λ, it holds that

$$0 = \liminf_{T\to\infty}\frac{1}{T}\mathbb{E}\left\{\lambda\left(x(T)\right)-\lambda\left(x(0)\right)\Big|\, x(0)\right\}$$

$$= \liminf_{T\to\infty}\frac{1}{T}\mathbb{E}\left\{\sum_{t=0}^{T-1}\mathbb{E}\left\{\lambda\left(x(t+1)\right)\Big|\, x(t)\right\}-\lambda\left(x(t)\right)\Big|\, x(0)\right\} \tag{A.51}$$

$$\le \liminf_{T\to\infty}\frac{1}{T}\mathbb{E}\left\{\sum_{t=0}^{T-1}s\left(x(t), u(t)\right)\Big|\, x(0)\right\}.$$

The second equality follows by the law of iterated expectation as T is fixed, the inequality results directly from Definition 5.13. Using the definition of the supply rate in (5.43) and taking into account that $\ell_\infty^{\mathrm{int}}(z_s^\infty, v_s^\infty)$ is a constant, we can derive

$$\liminf_{T\to\infty} \frac{1}{T} \mathbb{E}\left\{ \sum_{t=0}^{T-1} s\big(x(t), u(t)\big) \Big| x(0) \right\} \leq \liminf_{T\to\infty} \frac{1}{T} \sum_{t=0}^{T-1} \mathbb{E}\left\{ \ell\big(x(t), u(t)\big) \Big| x(0) \right\} - \ell_\infty^{\mathrm{int}}(z_s^\infty, v_s^\infty).$$
(A.52)

Combining the statements in (A.51) and (A.52) as well as recalling the input parametrization (2.21), stochastic optimal operation at steady-state is proven. □

A.12 Proof of Lemma 5.15

The stochastic available storage $S_{\mathrm{a}}(x)$ is Riemann integrable on $\overline{\mathbb{X}}_0^o$ if and only if it is bounded and continuous almost everywhere on $\overline{\mathbb{X}}_0^o$ (according to Lebesgue's criterion for Riemann integrability, see e.g., Zorich (2004, Section 11.1)).

By assumption, we know that the stochastic available storage is bounded, that is, $S_{\mathrm{a}}(x) < c < \infty$, and that $\mathbb{Z}$ is compact and convex. It remains to prove continuity almost everywhere on the domain.

In order to prove continuity almost everywhere, we introduce the following function representing the available storage for fixed T, i.e.,

$$S_{\mathrm{a}}^T(x) = \sup_{\substack{z(0)=x,\, z(k+1)=Az(k)+Bv(k),\\ (z(k),v(k))\in\overline{\mathbb{Z}}_k^o,\,\forall k\in\mathbb{I}_{\geq 0}}} \sum_{k=0}^{T-1} -\left(\ell_k^{\mathrm{int}}\big(z(k), v(k)\big) - \ell_\infty^{\mathrm{int}}(z_s, v_s) \right). \tag{A.53}$$

Moreover, we can define the set-valued map

$$\mathcal{U}_T(x) := \left\{ \boldsymbol{v} = \{v(0), \ldots, v(T-1)\} : z(0) = x,\, z(k+1) = Az(k) + Bv(k), \right.$$
$$\left. (z(k), v(k)) \in \overline{\mathbb{Z}}_k^o,\, \forall k \in \mathbb{I}_{[0,T-1]} \right\}. \tag{A.54}$$

Due to the preservation of convexity under Pontryagin difference with any arbitrary set (see, e.g., Kolmanovsky and Gilbert (1998, Theorem 2.1)), the constraints are all convex. Hence, following Grimm et al. (2004, Proposition 11&12) (for the case when $\mathbb{Z}$ is a polyhedron in $\mathbb{X} \times \mathbb{U}$ where $\mathbb{U}$ is a polytope, see also Rawlings and Mayne (2009, Section C.3)), the set valued map $\mathcal{U}_T$ is continuous on $\mathrm{int}(\overline{\mathbb{X}}_0^o)$. Moreover, since the terms ℓ_k^{int} are continuous for every $k \in \mathbb{I}_{[0,T-1]}$, the cost in (A.53) is continuous. With this, $S_{\mathrm{a}}^T(x)$ is continuous on $\mathrm{int}(\overline{\mathbb{X}}_0^o)$ as well.

We can state the following relation of (5.45) and (A.53) for each $x \in \overline{\mathbb{X}}_0^o$

$$\sup_{t\geq 0} S_{\mathrm{a}}^t(x) = S_{\mathrm{a}}(x), \tag{A.55}$$

i.e., the pointwise supremum of $S_{\mathrm{a}}^t(x)$ is equal to $S_{\mathrm{a}}(x)$. Using (A.53), we introduce

$$\overline{S}_{\mathrm{a}}^T(x) = \max_{T\geq t\geq 0} S_{\mathrm{a}}^t(x), \tag{A.56}$$

which shall be checked for uniform convergence to the limit $S_{\mathrm{a}}(x)$ on $\overline{\mathbb{X}}_0^o$ as $T \to \infty$.

As an intermediate step, we first investigate continuity of $\overline{S}_{\mathrm{a}}^{T}(x)$ by induction. We recall that $S_{\mathrm{a}}^{t}(x)$ is continuous for all $t \in \mathbb{I}_{\geq 0}$ on $\mathrm{int}(\overline{\mathbb{X}}_{0}^{o})$. As basis, $\overline{S}_{\mathrm{a}}^{1}(x) = \max\{S_{\mathrm{a}}^{0}(x), S_{\mathrm{a}}^{1}(x)\}$ is continuous, where the maximum is understood pointwise. Noting that $\overline{S}_{\mathrm{a}}^{k+1}(x) = \max\{\overline{S}_{\mathrm{a}}^{k}(x), S_{\mathrm{a}}^{k+1}(x)\}$ is continuous as well, this provides the inductive step and continuity of $\overline{S}_{\mathrm{a}}^{T}(x)$ for each finite $T \in \mathbb{I}_{\geq 0}$ and for all $x \in \mathrm{int}(\overline{\mathbb{X}}_{0}^{o})$.

Since continuity of $S_{\mathrm{a}}^{t}(x)$ is only provided for all $x \in \mathrm{int}(\overline{\mathbb{X}}_{0}^{o})$, we introduce an arbitrary compact subset of the interior of $\overline{\mathbb{X}}_{0}^{o}$ with non-empty interior denoted by $\mathcal{Y} \subset \mathrm{int}(\overline{\mathbb{X}}_{0}^{o})$. Since $\overline{\mathbb{X}}_{0}^{o}$ is a compact, convex set with non-empty interior (by Assumption 5.4), this is always possible.

Refocusing on uniform convergence, we define

$$a_{T} = \sup_{x \in \mathcal{Y}} \left| \overline{S}_{\mathrm{a}}^{T}(x) - S_{\mathrm{a}}(x) \right|. \tag{A.57}$$

Convergence of $\overline{S}_{\mathrm{a}}^{T}(x)$ uniformly on $\mathcal{Y}$ to the limit $S_{\mathrm{a}}(x)$ as $T \to \infty$ is equivalent to $a_{T} \to 0$ uniformly as $T \to \infty$. We can take two intermediate observations into account, (i) $\overline{S}_{\mathrm{a}}^{T}(x) \leq S_{\mathrm{a}}(x), \forall x \in \overline{\mathbb{X}}_{0}^{o}$ and $\forall T \in \mathbb{I}_{\geq 0}$, and (ii) $\overline{S}_{\mathrm{a}}^{T}(x) \leq \overline{S}_{\mathrm{a}}^{T+1}(x), \forall x \in \overline{\mathbb{X}}_{0}^{o}$ and $\forall T \in \mathbb{I}_{\geq 0}$. In order to provide convergence of a_{T}, investigate a_{T+1}

$$
\begin{aligned}
a_{T+1} &= \sup_{x \in \mathcal{Y}} \left| \overline{S}_{\mathrm{a}}^{T+1}(x) - S_{\mathrm{a}}(x) \right| \\
&\overset{(i)}{=} \sup_{x \in \mathcal{Y}} \left(S_{\mathrm{a}}(x) - \overline{S}_{\mathrm{a}}^{T+1}(x) \right) \\
&= \sup_{x \in \mathcal{Y}} \left(S_{\mathrm{a}}(x) - \overline{S}_{\mathrm{a}}^{T}(x) + \overline{S}_{\mathrm{a}}^{T}(x) - \overline{S}_{\mathrm{a}}^{T+1}(x) \right) \\
&\leq \underbrace{\sup_{x \in \mathcal{Y}} \left(S_{\mathrm{a}}(x) - \overline{S}_{\mathrm{a}}^{T}(x) \right)}_{=a_{T}} + \underbrace{\sup_{x' \in \mathcal{Y}} \left(\overline{S}_{\mathrm{a}}^{T}(x') - \overline{S}_{\mathrm{a}}^{T+1}(x') \right)}_{\leq 0, \text{ by observation (ii)}} \\
&\leq a_{T}
\end{aligned}
\tag{A.58}
$$

Taking the assumed boundedness of the available storage into account, i.e., $S_{\mathrm{a}}(x) < c < \infty$, we can see that $a_{T} < c$, $0 \leq a_{T}$, and by (A.58) that $a_{T+1} \leq a_{T}$. Hence, the series (a_{T}) converges as $T \to \infty$. It remains to show that it converges to the lower bound $\underline{a} = 0$.

We show uniform convergence on $\mathcal{Y}$ in the sense that $a_{T} \to 0$ as $T \to \infty$ by contradiction. Therefore, assume that $a_{T} \geq \underline{a} > 0$ for all $T \in \mathbb{I}_{\geq 0}$, that is, a_{T} is lower bounded by some $\underline{a} > 0$. Given any $x \in \mathcal{Y}$, for all r with $0 \leq r < S_{\mathrm{a}}(x)$ there exists a state/input sequence pair (z_{r}, v_{r}) starting at $z_{r}(0) = x$ and a $T_{r} \geq 0$ such that

$$\sum_{k=0}^{T_{r}-1} - \left(\ell_{k}^{\mathrm{int}}\left(z_{r}(k), v_{r}(k)\right) - \ell_{\infty}^{\mathrm{int}}(z_{s}, v_{s}) \right) > r. \tag{A.59}$$

We choose $r = S_{\mathrm{a}}(x) - \underline{a}/2$ and consider some $\bar{x} \in \mathcal{B}_{\epsilon}(x)$. Recalling continuity of $\mathcal{U}_{T}(x)$ on $\mathrm{int}(\overline{\mathbb{X}}_{0}^{o})$ and of the sum in (A.59), there exists an $\epsilon_{1} > 0$ and a feasible input sequence $\bar{v}_{r}$ such that for all $\bar{x} \in \mathcal{B}_{\epsilon_{1}}(x)$, with $\bar{z}_{r}(0) = \bar{x}$,

$$\sum_{k=0}^{T_{r}-1} - \left(\ell_{k}^{\mathrm{int}}\left(\bar{z}_{r}(k), \bar{v}_{r}(k)\right) - \ell_{\infty}^{\mathrm{int}}(z_{s}, v_{s}) \right) > r. \tag{A.60}$$

This leads to

$$\left|\bar{S}_a^{T_r}(y) - S_a(y)\right| < \underline{a}/2, \ \forall y \in \mathcal{B}_{\epsilon_1}(x). \tag{A.61}$$

By the definition of compact sets (Munkres, 2000), we can find a finite subcover of $\mathcal{Y}$ consisting of finitely many open balls $\mathcal{B}_{\epsilon_{1,i}}(x_i)$ such that $\bigcup_i \mathcal{B}_{\epsilon_{1,i}}(x_i) \supseteq \mathcal{Y}$. The x_i and $\epsilon_{1,i}$ are chosen such that we can find a state/input sequence pair $(\boldsymbol{z}_{r,i}, \boldsymbol{v}_{r,i})$ and a $T_{r,i} \geq 0$ with

$$\left|\bar{S}_a^{T_{r,i}}(y_i) - S_a(y_i)\right| \leq \underline{a}/2, \ \forall y_i \in \mathcal{B}_{\epsilon_{1,i}}(x_i). \tag{A.62}$$

Since (A.59) holds for any $x \in \mathcal{Y}$, there always exists such an open ball for any $x_i \in \mathcal{Y}$. The construction of the finite subcover can be conducted similar to the procedure described in Khalil (2002, Exercise 3.19). We note that this procedure is only possible for $\mathcal{Y}$ and not for $\overline{\mathbb{X}}_0^o$, since continuity of the set valued map is only provided on $\mathrm{int}(\overline{\mathbb{X}}_0^o)$ and a finite subcover can only be found for compact sets. With $T^* = \max_i T_{r,i}$,

$$\left|\bar{S}_a^{T^*}(y) - S_a(y)\right| \leq \underline{a}/2, \ \forall y \in \mathcal{Y}, \tag{A.63}$$

and thus, the desired contradiction follows: For each assumed lower limit $\underline{a} > 0$, there can always be found a smaller lower limit, and hence, $\underline{a}$ must be zero.

Now, we can employ the *uniform limit theorem* (Munkres, 2000, Theorem 21.6) stating that if a sequence of functions f_n converges uniformly to a limit function f and if each of the functions f_n is continuous, then the limit f must be continuous as well.

Since we have satisfied all preliminaries for the uniform limit theorem—$\overline{S}_a^T(x)$ is continuous for all $T \in \mathbb{I}_{\geq 0}$ (in $\mathcal{Y}$) and $\overline{S}_a^T(x)$ converges uniformly to $S_a(x)$ on $\mathcal{Y}$ as $T \to \infty$—the stochastic available storage $S_a(x)$ is continuous on $\mathcal{Y}$.

It remains to show that the available storage $S_a(x)$ is continuous almost everywhere on $\overline{\mathbb{X}}_0^o$. Since $\mathcal{Y}$ can be any arbitrary compact set in the interior of $\overline{\mathbb{X}}_0^o$, for all $\bar{x} \in \mathrm{int}(\overline{\mathbb{X}}_0^o)$, there exists a compact set $\overline{\mathcal{Y}} \subset \mathrm{int}(\overline{\mathbb{X}}_0^o)$ with $\bar{x} \in \mathrm{int}(\overline{\mathcal{Y}})$. Hence, we can derive continuity for all $x \in \mathrm{int}(\overline{\mathbb{X}}_0^o)$, and only on the boundary of $\overline{\mathbb{X}}_0^o$ continuity of $S_a(x)$ is not guaranteed. However, by this we have continuity of $S_a(x)$ almost everywhere on $\overline{\mathbb{X}}_0^o$. $\qquad\square$

A.13 Proof of Lemma 5.16

The proof follows along the lines of (Berman and Shaked, 2006; Willems, 1972; Wu et al., 2011). For ease of presentation, we assume, without loss of generality, that $\ell_\infty^{\mathrm{int}}(z_s, v_s) = 0$.

To start with, we consider sufficiency and assume that there exists a constant $c < \infty$ such that $S_a(x(0)) < c$ for all $x(0) \in \overline{\mathbb{X}}_0^o$. Therefore, one has to show that $S_a(x)$ is an appropriate storage function. For any $T \geq T' \geq 0$, it holds that

$$
\begin{aligned}
S_a\big(x(0)\big) &= \sup_{\substack{T \geq 0, \\ z(0)=x(0),\, z(k+1)=Az(k)+Bv(k), \\ (z(k),v(k)) \in \overline{\mathbb{Z}}_k^o,\, \forall k \in \mathbb{I}_{\geq 0}}} \sum_{k=0}^{T-1} -\ell_k^{\mathrm{int}}\big(z(k), v(k)\big) \\[2mm]
&\geq -\sum_{k=0}^{T'-1} \ell_k^{\mathrm{int}}\big(z(k), v(k)\big) + \sup_{\substack{T \geq T', \\ z(T')=x',\, z(k+1)=Az(k)+Bv(k), \\ (z(k),v(k)) \in \overline{\mathbb{Z}}_k^o,\, \forall k \in \mathbb{I}_{\geq T'}}} \sum_{k=T'}^{T-1} -\ell_k^{\mathrm{int}}\big(z(k), v(k)\big)
\end{aligned}
\tag{A.64}
$$

for all feasible trajectories $(z(k), v(k)) \in \overline{\mathbb{Z}}_k^o$, $k \in \mathbb{I}_{[0,T'-1]}$, using the abbreviation $x' = A^{T'} x(0) + \sum_{k=0}^{T'-1} A^{T'-k-1} Bv(k)$.

Moreover, one can see that

$$
\sup_{\substack{T \geq T', \\ z(T')=x', z(k+1)=Az(k)+Bv(k), \\ (z(k),v(k))\in\overline{\mathbb{Z}}_k^o, \forall k\in\mathbb{I}_{\geq T'}}} \sum_{k=T'}^{T-1} -\ell_k^{\text{int}}\Big(z(k), v(k)\Big)
$$

$$
= \mathbb{E}\left\{ \sup_{\substack{T \geq 0, \\ z(0)=x(T'), z(k+1)=Az(k)+Bv(k), \\ (z(k),v(k))\in\overline{\mathbb{Z}}_k^o, \forall k\in\mathbb{I}_{\geq 0}}} \sum_{k=0}^{T-1} -\ell_k^{\text{int}}\Big(z(k), v(k)\Big) \middle| x(0) \right\} = \mathbb{E}\left\{ S_{\text{a}}\Big(x(T')\Big) \middle| x(0) \right\}.
$$

(A.65)

The equality follows by $x(T') = x' + \sum_{k=0}^{T'-1} A_{\text{cl}}^{T'-k-1} w(k)$, where $w(k) \in \mathbb{W}$, $k \in \mathbb{I}_{[0,T'-1]}$, with PDF $\rho_\mathbb{W}$ (see Assumption 2.12). We note that by Riemann integrability provided in Lemma 5.15, the conditional expectation of the available storage in (A.65) exists.

With (A.65) and setting $T' = 1$, we can rewrite (A.64) such that

$$
S_{\text{a}}(x(0)) \geq -\ell(x(0), u(0)) + \mathbb{E}\left\{ S_{\text{a}}\Big(x(1)\Big) \middle| x(0) \right\}, \tag{A.66}
$$

which provides stochastic dissipativity with the storage function $S_{\text{a}}(x)$.

Next, we turn to necessity and assume that the system is dissipative with respect to some locally integrable and bounded storage function. Thus, by dissipativity (using similar transformations as in (A.64) and (A.65)), it follows that

$$
\lambda\Big(x(0)\Big) + \sum_{k=0}^{T-1} \ell_k^{\text{int}}\Big(z(k), v(k)\Big) \geq \mathbb{E}\left\{ \lambda\Big(x(T)\Big) \middle| x(0) \right\} \geq 0, \tag{A.67}
$$

which shows that

$$
\lambda\Big(x(0)\Big) \geq \sup_{\substack{T \geq 0, \\ z(0)=x(0), z(k+1)=Az(k)+Bv(k), \\ (z(k),v(k))\in\overline{\mathbb{Z}}_k^o, \forall k\in\mathbb{I}_{\geq 0}}} \sum_{k=0}^{T-1} -\ell_k^{\text{int}}\Big(z(k), v(k)\Big) = S_{\text{a}}\Big(x(0)\Big). \tag{A.68}
$$

Hence, due to the storage function λ being bounded on $\overline{\mathbb{X}}_0^o$, there exists a constant c such that

$$
S_{\text{a}}\Big(x(0)\Big) \leq \lambda\Big(x(0)\Big) < c < \infty, \tag{A.69}
$$

which closes the proof. $\square$

A.14 Proof of Theorem 5.20

The proof idea follows from Müller et al. (2015a) and is based on proof by contradiction. Therefore, we assume the system to be stochastically optimally operated at steady-state, that is, for each feasible nominal state/input sequence pair (z, v), it holds that

$$
\liminf_{T \to \infty} \frac{1}{T} \sum_{t=0}^{T-1} \left(\mathbb{E} \left\{ \ell\big(x(t), u(t)\big) \,\big|\, x(0) \right\} - \ell_\infty^{\text{int}}(z_s^\infty, v_s^\infty) \right)
$$
$$
= \liminf_{T \to \infty} \frac{1}{T} \sum_{t=0}^{T-1} \left(\ell_t^{\text{int}}\big(x(t), u(t)\big) - \ell_\infty^{\text{int}}(z_s^\infty, v_s^\infty) \right) \geq 0
$$

(A.70)

In addition, we assume that the system is *not* stochastically dissipative on $\overline{\mathcal{Z}}_M$. By Lemma 5.16 (and taking Remark 5.19 into account), this is equivalent to an unbounded available storage on $\overline{\mathcal{X}}_M$, i.e., for each $r \geq 0$ there exists a $y \in \overline{\mathcal{X}}_M$ such that

$$
\inf_{\substack{T \geq 0, \\ z(0) = y,\, z(k+1) = Az(k) + Bv(k), \\ (z(k), v(k)) \in \overline{\mathcal{Z}}_M,\, \forall k \in \mathbb{I}_{\geq 0}}} \sum_{k=0}^{T-1} \left(\ell_k^{\text{int}}\big(z(k), v(k)\big) - \ell_\infty^{\text{int}}(z_s^\infty, v_s^\infty) \right) \leq -r.
$$

(A.71)

This means that for each $r \geq 0$ there must exist an admissible nominal state/input sequence pair (z_r, v_r) (satisfying the nominal dynamics) with $z_r(0) = y$ and $(z_r(k), v_r(k)) \in \overline{\mathcal{Z}}_M$ for all $k \in \mathbb{I}_{\geq 0}$ together with $T_r \geq 0$ and

$$
\sum_{k=0}^{T_r-1} \left(\ell_k^{\text{int}}\big(z_r(k), v_r(k)\big) - \ell_\infty^{\text{int}}(z_s^\infty, v_s^\infty) \right) \leq -r.
$$

(A.72)

In the following discussion, we fix r to be

$$
r = 1 + c + 2M\ell_{\max} - 2M\ell_\infty^{\text{int}}(z_s^\infty, v_s^\infty),
$$

(A.73)

where c is the constant introduced in Assumption 5.5 and $\ell_{\max} = \max_{(z,v) \in \overline{\mathcal{Z}}_{2M}} \ell_\infty^{\text{int}}(z, v)$. Due to boundedness of ℓ on $\overline{\mathcal{Z}}_{2M}$, r chosen according to (A.73) is bounded. Using (A.72) and Assumption 5.5, one can show that

$$
\sum_{k=0}^{T_r-1} \left(\ell_\infty^{\text{int}}\big(z_r(k), v_r(k)\big) - \ell_\infty^{\text{int}}(z_s^\infty, v_s^\infty) \right) \leq -r + c.
$$

(A.74)

Following the approach in Müller et al. (2015a), by definition, $z_r(\cdot) \in \overline{\mathcal{C}}_M \cap \overline{\mathcal{R}}_M$ such that there exists a second nominal state/input sequence pair (z', v') with $z'(0) = z_r(T_r)$, $z'(M) = z_s^\infty$, and $z'(2M) = y$ as well as $(z'(k), v'(k)) \in \overline{\mathcal{Z}}_\infty$ for all $k \in \mathbb{I}_{[0,2M]}$. Moreover, since $z'(0) = z_r(T_r) \in \overline{\mathcal{X}}_M$ and $z'(2M) = y \in \overline{\mathcal{X}}_M$, it holds that $(z'(k), v'(k)) \in \overline{\mathcal{Z}}_{2M}$ for all $k \in \mathbb{I}_{[0,2M]}$. Defining the input sequence

$$
\hat{v}\big(t(T_r + 2M) + i\big) = \begin{cases} v_r(i), & t \in \mathbb{I}_{\geq 0}, i \in \mathbb{I}_{[0, T_r-1]}, \\ v'(i - T_r), & t \in \mathbb{I}_{\geq 0}, i \in \mathbb{I}_{[T_r, T_r+2M-1]}, \end{cases}
$$

(A.75)

results in a cyclic (and feasible) nominal state sequence $\hat{z}$ satisfying $\hat{z}(t(T_r + 2M)) = y$ and $\hat{z}(t(T_r + 2M) + T_r + M) = z_s^\infty$ for all $t \in \mathbb{I}_{\geq 0}$. Thus, it follows directly that $(\hat{z}(k), \hat{v}(k)) \in \overline{\mathcal{Z}}_{2M} \subseteq \overline{\mathbb{Z}}_\infty^o$ for all $k \in \mathbb{I}_{\geq 0}$. Evaluating ℓ_∞^{int} along one cycle of $(\hat{z}, \hat{v})$ of length $T_r + 2M$ leads to

$$\sum_{i=0}^{T_r+2M-1} \ell_\infty^{\text{int}}\big(\hat{z}(i), \hat{v}(i)\big) = \sum_{i=0}^{T_r-1} \ell_\infty^{\text{int}}\big(z_r(i), v_r(i)\big) + \sum_{i=0}^{2M-1} \ell_\infty^{\text{int}}\big(z'(i), v'(i)\big)$$

$$\leq -r + c + T_r\,\ell_\infty^{\text{int}}(z_s^\infty, v_s^\infty) + \sum_{i=0}^{2M-1} \ell_\infty^{\text{int}}\big(z'(i), v'(i)\big) \tag{A.76}$$

$$\leq -r + c + T_r\,\ell_\infty^{\text{int}}(z_s^\infty, v_s^\infty) + 2M\ell_{\max},$$

where the first inequality follows from (A.74).

With these preliminary statements, we can derive for the average of the expected value of the cost along $(\hat{z}, \hat{v})$ as $T \to \infty$

$$\liminf_{T\to\infty} \frac{1}{T} \sum_{k=0}^{T-1} \ell_k^{\text{int}}\big(\hat{z}(k), \hat{v}(k)\big)$$

$$= \liminf_{T\to\infty} \frac{1}{T} \sum_{k=0}^{T-1} \Big(\ell_k^{\text{int}}\big(\hat{z}(k), \hat{v}(k)\big) + \ell_\infty^{\text{int}}\big(\hat{z}(k), \hat{v}(k)\big) - \ell_\infty^{\text{int}}\big(\hat{z}(k), \hat{v}(k)\big)\Big)$$

$$\leq \liminf_{T\to\infty} \frac{1}{T} \Big(c + \sum_{k=0}^{T-1} \ell_\infty^{\text{int}}\big(\hat{z}(k), \hat{v}(k)\big)\Big)$$

$$\leq \limsup_{T\to\infty} \frac{1}{T} c + \frac{1}{T_r+2M} \sum_{k=0}^{T_r+2M-1} \ell_\infty^{\text{int}}\big(\hat{z}(k), \hat{v}(k)\big) \tag{A.77}$$

$$\leq \frac{1}{T_r+2M}\Big(-r + c + T_r\,\ell_\infty^{\text{int}}(z_s^\infty, v_s^\infty) + 2M\ell_{\max}\Big)$$

$$= \frac{1}{T_r+2M}\Big(-1 + (T_r+2M)\,\ell_\infty^{\text{int}}(z_s^\infty, v_s^\infty)\Big)$$

$$< \ell_\infty^{\text{int}}(z_s^\infty, v_s^\infty),$$

where the second equality follows with the choice of r in (A.73). Obviously, this contradicts the definition of stochastic optimal operation at steady-state (see (A.70)). $\qquad\square$

A.15 Proof of Proposition 5.21

The proof idea follows along the lines of the approach in Butzer and Hahn (1978). We present the proof for the scalar case only, i.e., $\ell : \mathbb{R} \times \mathbb{R} \to \mathbb{R}$, in order to simplify the presentation. However, it also applies for the general case $\ell : \mathbb{R}^n \times \mathbb{R}^m \to \mathbb{R}$.

Since polynomial stage cost functions of finite degree are considered, the integrated stage cost ℓ_k^{int} is a Taylor series of finite order. Recalling the definition of ℓ_k^{int} in (5.11), writing this function by its integral form, and fixing z and v, we consider the Taylor series of $\ell(z + \epsilon, v + K\epsilon)$ with respect to ϵ at zero. Thus, the integrated stage cost can be rewritten by

$$\ell_k^{\text{int}}(z, v) = \sum_{j=0}^{\bar{r}} \frac{1}{j!} \ell^{(j)}(z, v) \int_{\mathbb{R}} \epsilon^j \rho_{\Omega_k}(\epsilon)\,\mathrm{d}\epsilon, \tag{A.78}$$

where $\ell^{(j)}(z, v)$ is the j-th derivative of $\ell(z + \epsilon, v + K\epsilon)$ with respect to ϵ evaluated at $\epsilon = 0$. We note that the factor before the integral is deterministic, i.e., independent of ϵ, while the integral is the j-th moment of $\rho_{\Omega_k}(\epsilon)$. Due to boundedness of Ω_k, one can bound the integral from above by $c \int_{\mathbb{R}} \rho_{\Omega_k}(\epsilon) \, d\epsilon$, with $\max_{\epsilon \in \Omega_k} \epsilon^j \leq c < \infty$, and thus, all moments are bounded as well. In addition, all derivatives $\ell^{(p)}$ with $p \geq \bar{r} + 1$ are equal to zero since we consider polynomial functions only. Using (A.78), we can rewrite the single terms in (5.49) by

$$\ell_k^{\mathrm{int}}(z,v) - \ell_\infty^{\mathrm{int}}(z,v) = \sum_{j=0}^{\bar{r}} \frac{1}{j!} \ell^{(j)}(z,v) \left(\int_{\mathbb{R}} \epsilon^j \, \rho_{\Omega_k}(\epsilon) \, d\epsilon - \int_{\mathbb{R}} \epsilon^j \, \rho_{\Omega_\infty}(\epsilon) \, d\epsilon \right). \tag{A.79}$$

Similar as used in the proof of Lemma 5.2, each $\epsilon \in \Omega_k$ can be represented by $\sum_{i=0}^{k-1} A_{\mathrm{cl}}^i w_i$ with $w_i \in \mathbb{W}$, recalling that all disturbances are i.i.d. and have zero mean. Using Fubini's theorem and the binomial theorem, it follows that all terms which multiply with A_{cl}^m, $m \in \mathbb{I}_{[0,k-1]}$ cancel, and hence

$$\left| \ell_k^{\mathrm{int}}(z,v) - \ell_\infty^{\mathrm{int}}(z,v) \right| = O\left(A_{\mathrm{cl}}^k \right). \tag{A.80}$$

Since the eigenvalues of A_{cl} are strictly inside the unit disc, the error sum in (5.49) is a geometric series, and thus, bounded. □

Remark A.1. *As noted in the proof, when the arguments of the stage cost function are not scalar, i.e., $\ell : \mathbb{R}^n \times \mathbb{R}^m \to \mathbb{R}$, one can use the same approach. We note that the derivatives become multidimensional tensors.*

If cost functions are considered that are not polynomial functions of finite degree, one can still use the same idea of a Taylor series. However, then a bound on the residual term is needed (see Butzer and Hahn (1978)) in order to account for the neglected higher order terms such that convergence for the sum of errors is not guaranteed and the computation needed in order to prove convergence can become more involved.

A.16 Proof of Theorem 6.4

The proof follows along the lines of the proof of Angeli et al. (2012, Proposition 6.4). Due to λ being bounded from below, we can see that

$$\begin{aligned}
0 &\leq \liminf_{T \to \infty} \frac{1}{T} \left(\lambda\big(x(T)\big) - \lambda\big(x(0)\big) \right) \\
&\leq \liminf_{T \to \infty} \frac{1}{T} \sum_{t=0}^{T-1} \left(s\big(x(t), u(t)\big) + \rho(w_{\mathrm{max}}) \right) \\
&= \liminf_{T \to \infty} \frac{1}{T} \sum_{t=0}^{T-1} \ell\big(x(t), u(t)\big) - \ell\big(x_s(w_{\mathrm{max}}), u_s(w_{\mathrm{max}})\big) + \rho(w_{\mathrm{max}}),
\end{aligned} \tag{A.81}$$

where the second inequality follows from (6.3) and the last equality from (6.6). By reordering (A.81), $\mathcal{K}$-robust optimal operation at steady-state according to (6.4) follows. □

A.17 Proof of Lemma 6.6

Since $\mathbb{W}(w_{\mathrm{max}})$ is a compact and convex set containing the origin in its interior, it holds that $\mathbb{W}(w_{\mathrm{max}}) \subseteq w_{\mathrm{max}} \overline{\mathcal{B}}_1$. It follows for the mRPI set of the associated error dynamics

$$\Omega(w_{\mathrm{max}}) \subseteq w_{\mathrm{max}} \bigoplus_{i=0}^{\infty} A^i \overline{\mathcal{B}}_1 := w_{\mathrm{max}} \overline{\Omega}, \tag{A.82}$$

where $\overline{\Omega}$ exists since A is asymptotically stable and $\overline{\mathcal{B}}_1$ is bounded. The tightened set $\overline{\mathbb{Z}}_s(w_{\max})$, which provides all feasible steady-states, is given by

$$\overline{\mathbb{Z}}_s(w_{\max}) = \left\{ (x, u) \in \mathbb{R}^n \times \mathbb{R}^m : x = Ax + Bu, (x, u) \in \mathbb{Z} \ominus \left(\Omega(w_{\max}) \times 0 \right) \right\}, \quad (A.83)$$

This can be understood as a set-valued function $w_{\max} \mapsto \overline{\mathbb{Z}}_s(w_{\max})$. Considering the constraint tightening in (A.83), it holds

$$\mathbb{Z} \ominus \left(w_{\max} \overline{\Omega} \times 0 \right) \subseteq \mathbb{Z} \ominus \left(\Omega(w_{\max}) \times 0 \right) \subseteq \mathbb{Z}, \quad (A.84)$$

and thus, $\mathbb{Z} \ominus \left(\Omega(w_{\max}) \times 0 \right) \to \mathbb{Z}$ as $w_{\max} \to 0$. With this relation, continuity of $\overline{\mathbb{Z}}_s(w_{\max})$ at $w_{max} = 0$ can be derived by the definition of continuity of set-valued functions, see, e.g., Rockafellar and Wets (2009, Definition 5.4). We recall that ℓ is continuous by Assumption 2.3 and $\overline{\mathbb{Z}}_s(w_{\max})$ is compact-valued (by Assumption 2.2 and Kolmanovsky and Gilbert (1998, Theorem 2.1)). Hence, by Rawlings and Mayne (2009, Theorem C.28), $\ell(x_s(\cdot), u_s(\cdot))$ is continuous (and $(x_s(\cdot), u_s(\cdot))$ is outer semicontinuous) at $w_{\max} = 0$. $\quad \square$

Bibliography

R. P. Aguilera and D. E. Quevedo. Stability analysis of quadratic MPC with a discrete input alphabet. *IEEE Transactions on Automatic Control*, 58(12):3190–3196, 2013.

D. A. Allan, C. N. Bates, M. J. Risbeck, and J. B. Rawlings. On the inherent robustness of optimal and suboptimal MPC. Technical Report 2016-01, TWCCC, 2016. URL http://jbrwww.che.wisc.edu/tech-reports/twccc-2016-01.pdf.

B. Alrifaee, Y. Liu, and D. Abel. ECO-cruise control using economic model predictive control. In *Proceedings of the IEEE Conference on Control Applications*, pages 1933–1938, 2015.

R. Amrit, J. B. Rawlings, and D. Angeli. Economic optimization using model predictive control with a terminal cost. *Annual Reviews in Control*, 35(2):178–186, 2011.

D. Angeli and J. B. Rawlings. Receding horizon cost optimization and control for nonlinear plants. In *Proceedings of the 8th IFAC Symposium on Nonlinear Control Systems*, pages 1217–1223, 2010.

D. Angeli, R. Amrit, and J. B. Rawlings. Receding horizon cost optimization for overly constrained nonlinear plants. In *Proceedings of the 48th IEEE Conference on Decision and Control, held jointly with the 28th Chinese Control Conference*, pages 7972–7977, 2009.

D. Angeli, R. Amrit, and J. B. Rawlings. Enforcing convergence in nonlinear economic MPC. In *Proceedings of the 50th IEEE Conference on Decision and Control and European Control Conference*, pages 3387–3391, 2011.

D. Angeli, R. Amrit, and J. B. Rawlings. On average performance and stability of economic model predictive control. *IEEE Transactions on Automatic Control*, 57(7):1615–1626, 2012.

D. Angeli, A. Casavola, and F. Tedesco. Theoretical advances on economic model predictive control with time-varying costs. *Annual Reviews in Control*, 41:218–224, 2016.

T. Backx, O. Bosgra, and W. Marquardt. Integration of model predictive control and optimization of processes. In *Proceedings of the 4th IFAC Symposium on Advanced Control of Chemical Processes*, pages 249–260, 2000.

F. A. Bayer and F. Allgöwer. Robust economic model predictive control with linear average constraints. In *Proceedings of the 53rd IEEE Conference on Decision and Control*, pages 6707–6712, 2014.

F. A. Bayer, M. Bürger, and F. Allgöwer. Discrete-time incremental ISS: A framework for robust NMPC. In *Proceedings of the European Control Conference*, pages 2068–2073, 2013.

F. A. Bayer, M. A. Müller, and F. Allgöwer. Set-based disturbance attenuation in economic model predictive control. In *Proceedings of the 19th IFAC World Congress*, pages 1898–1903, 2014a.

F. A. Bayer, M. A. Müller, and F. Allgöwer. Tube-based robust economic model predictive control. *Journal of Process Control*, 24(8):1237–1246, 2014b.

F. A. Bayer, M. Lorenzen, M. A. Müller, and F. Allgöwer. Improving performance in robust economic MPC using stochastic information. In *Proceedings of the 5th IFAC Conference on Nonlinear Model Predictive Control*, pages 411–416, 2015a.

F. A. Bayer, M. A. Müller, and F. Allgöwer. Average constraints in robust economic model predictive control. In *Proceedings of the 9th IFAC Symposium on Advanced Control of Chemical Processes*, pages 44–49, 2015b.

F. A. Bayer, M. Lorenzen, M. A. Müller, and F. Allgöwer. Robust economic model predictive control using stochastic information. *Automatica*, 74:151–161, 2016a.

F. A. Bayer, M. A. Müller, and F. Allgöwer. Min-max economic model predictive control approaches with guaranteed performance. In *Proceedings of the 55th IEEE Conference on Decision and Control*, pages 3210–3215, 2016b.

F. A. Bayer, M. A. Müller, and F. Allgöwer. On optimal operation in robust economic MPC. *Automatica*, 2017. to appear.

A. Bemporad and M. Morari. Robust model predictive control: A survey. In A. Garulli and A. Tesi, editors, *Robustness in Identification and Control*, pages 207–226. Springer London, 1999.

A. Bemporad, F. Borrelli, and M. Morari. Min-max control of constrained uncertain discrete-time linear systems. *IEEE Transactions on Automatic Control*, 48(9):1600–1606, 2003.

A. Ben-Tal, L. E. Ghaoui, and A. Nemirovski. *Robust Optimization*. Princeton University Press, 2009.

N. Berman and U. Shaked. H_∞-control for discrete-time nonlinear stochastic systems. *IEEE Transactions on Automatic Control*, 51(6):1041–1046, 2006.

D. Bernardini and A. Bemporad. Stabilizing model predictive control of stochastic constrained linear systems. *IEEE Transactions on Automatic Control*, 57(6):1468–1480, 2012.

D. P. Bertsekas. *Dynamic Programming and Optimal Control*, volume 1. Athena Scientific, second edition, 2005.

D. Bertsimas and M. Sim. The price of robustness. *Operations Research*, 52(1):35–53, 2004.

D. Bertsimas, D. Pachamanova, and M. Sim. Robust linear optimization under general norms. *Operations Research Letters*, 32(6):510–516, 2004.

D. Bertsimas, D. B. Brown, and C. Caramanis. Theory and applications of robust optimization. *SIAM Review*, 53(3):464–501, 2011.

H.-G. Beyer and B. Sendhoff. Robust optimization – A comprehensive survey. *Computer Methods in Applied Mechanics and Engineering*, 196(33–34):3190–3218, 2007.

L. Blackmore, M. Ono, A. Bektassov, and B. C. Williams. A probabilistic particle-control approximation of chance-constrained stochastic predictive control. *IEEE Transactions on Robotics*, 26(3):502–517, 2010.

F. Blanchini. Set invariance in control. *Automatica*, 35(11):1747–1767, 1999.

T. Bø and T. A. Johansen. Dynamic safety constraints by scenario based economic model predictive control of marine electric power plants. *IEEE Transactions on Transportation Electrification*, 3(1):13–21, 2017.

V. S. Borkar. Controlled diffusion processes. *Probability surveys*, 2(4):213–244, 2005.

V. S. Borkar and M. K. Ghosh. Ergodic control of multidimensional diffusions I: The existence results. *SIAM Journal on Control and Optimization*, 26(1):112–126, 1988.

S. Boyd and L. Vandenberghe. *Convex Optimization*. Cambridge University Press, 2004.

T. J. Broomhead, C. Manzie, R. Shekhar, M. Brear, and P. Hield. Robust stable economic MPC with applications in engine control. In *Proceedings of the 53rd IEEE Conference on Decision and Control*, pages 2511–2516, 2014.

T. J. Broomhead, C. Manzie, R. C. Shekhar, and P. Hield. Robust periodic economic MPC for linear systems. *Automatica*, 60:30–37, 2015.

T. J. Broomhead, C. Manzie, P. Hield, R. Shekhar, and M. Brear. Economic model predictive control and applications for diesel generators. *IEEE Transactions on Control Systems Technology*, 25(2):388–400, 2017.

F. D. Brunner and F. Allgöwer. A Lyapunov function approach to the event-triggered stabilization of the minimal robust positively invariant set. In *Proceedings of the 5th IFAC Workshop on Distributed Estimation and Control in Networked System*, pages 25–30, 2015.

F. D. Brunner, F. A. Bayer, and F. Allgöwer. Robust steady state optimization for polytopic systems. In *Proceedings of the 55th IEEE Conference on Decision and Control*, pages 4084–4089, 2016.

H.-J. Bungartz and M. Griebel. Sparse grids. *Acta Numerica*, 13:147–269, 2004.

P. L. Butzer and L. Hahn. General theorems on rates of convergence in distribution of random variables I. General limit theorems. *Journal of Multivariate Analysis*, 8(2):181–201, 1978.

C. Byrnes and W. Lin. Losslessness, feedback equivalence, and the global stabilization of discrete-time nonlinear systems. *IEEE Transactions on Automatic Control*, 39(1):83–98, 1994.

P. J. Campo and M. Morari. Robust model predictive control. In *Proceedings of the American Control Conference*, pages 1021–1026, 1987.

M. Cannon, B. Kouvaritakis, and D. Ng. Probabilistic tubes in linear stochastic model predictive control. *Systems & Control Letters*, 58(10-11):747–753, 2009a.

M. Cannon, B. Kouvaritakis, and X. Wu. Probabilistic constrained MPC for multiplicative and additive stochastic uncertainty. *IEEE Transactions on Automatic Control*, 54(7): 1626–1632, 2009b.

M. Cannon, B. Kouvaritakis, S. V. Raković, and Q. Cheng. Stochastic tubes in model predictive control with probabilistic constraints. *IEEE Transactions on Automatic Control*, 56(1):194–200, 2011.

M. Cannon, Q. Cheng, B. Kouvaritakis, and S. V. Raković. Stochastic tube MPC with state estimation. *Automatica*, 48(3):536–541, 2012.

D. Chatterjee and J. Lygeros. On stability and performance of stochastic predictive control techniques. *IEEE Transactions on Automatic Control*, 60(2):509–514, 2015.

H. Chen and F. Allgöwer. A quasi-infinite horizon nonlinear model predictive control scheme with guaranteed stability. *Automatica*, 34(10):1205–1217, 1998.

H. Chen, C. W. Scherer, and F. Allgower. A game theoretic approach to nonlinear robust receding horizon control of constrained systems. In *Proceedings of the American Control Conference*, volume 5, pages 3073–3077, 1997.

L. Chisci and G. Zappa. Feasibility in predictive control of constrained linear systems: The output feedback case. *International Journal of Robust and Nonlinear Control*, 12(5): 465–487, 2002.

L. Chisci, J. A. Rossiter, and G. Zappa. Systems with persistent disturbances: Predictive control with restricted constraints. *Automatica*, 37(7):1019–1028, 2001.

B. Chu, S. Duncan, A. Papachristodoulou, and C. Hepburn. Using economic model predictive control to design sustainable policies for mitigating climate change. In *Proceedings of the 51st IEEE Conference on Decision and Control*, pages 406–411, 2012.

T. Damm, L. Grüne, M. Stieler, and K. Worthmann. An exponential turnpike theorem for dissipative discrete time optimal control problems. *SIAM Journal on Optimization*, 52 (3):1935–1957, 2014.

L. Deori, S. Garatti, and M. Prandini. Stochastic control with input and state constraints: A relaxation technique to ensure feasibility. In *Proceedings of the 54th IEEE Conference on Decision and Control*, pages 3786–3791, 2015.

S. Di Cairano. Indirect adaptive model predictive control for linear systems with polytopic uncertainty. In *Proceedings of the American Control Conference*, pages 3570–3575, 2016.

M. Diehl, R. Amrit, and J. B. Rawlings. A Lyapunov function for economic optimizing model predictive control. *IEEE Transactions on Automatic Control*, 56(3):703–707, 2011.

A. D'Jorge, A. Anderson, A. H. González, and A. Ferramosca. A robust economic MPC for changing economic criterion. In *Proceedings of the IEEE Conference on Control Applications*, pages 1374–1379, 2016.

M. Ellis, H. Durand, and P. D. Christofides. A tutorial review of economic model predictive control methods. *Journal of Process Control*, 24(8):1156–1178, 2014.

S. Engell. Feedback control for optimal process operation. *Journal of Process Control*, 17 (3):203–219, 2007.

L. Fagiano and M. Khammash. Nonlinear stochastic model predictive control via regularized polynomial chaos expansions. In *Proceedings of the 51st IEEE Conference on Decision and Control*, pages 142–147, 2012.

T. Faulwasser and D. Bonvin. On the design of economic NMPC based on approximate turnpike properties. In *Proceedings of the 54th IEEE Conference on Decision and Control*, pages 4964–4970, 2015.

T. Faulwasser and D. Bonvin. Exact turnpike properties and economic NMPC. *European Journal of Control*, 35:34–41, 2017.

A. Ferramosca, A. H. González, and D. Limón. Offset-free multi-model economic model predictive control for changing economic criterion. *Journal of Process Control*, 54:1–13, 2017.

R. Findeisen, L. Imsland, F. Allgöwer, and B. A. Foss. State and output feedback nonlinear model predictive control: An overview. *European Journal of Control*, 9:179–195, 2003.

F. A. Fontes. A general framework to design stabilizing nonlinear model predictive controllers. *Systems & Control Letters*, 42(2):127–143, 2001.

H. Genceli and M. Nikolaou. Robust stability analysis of constrained ℓ_1-norm model predictive control. *AIChE Journal*, 39(12):1954–1965, 1993.

E. Gilbert and K. T. Tan. Linear systems with state and control constraints: The theory and application of maximal output admissible sets. *IEEE Transactions on Automatic Control*, 36(9):1008–1020, 1991.

R. Goebel, R. G. Sanfelice, and A. R. Teel. *Hybrid Dynamical Systems: Modeling, Stability, and Robustness*. Princeton University Press, 2012.

P. J. Goulart and E. C. Kerrigan. Output feedback receding horizon control of constrained systems. *International Journal of Control*, 80(1):8–20, 2007.

P. J. Goulart, E. C. Kerrigan, and J. M. Maciejowski. Optimization over state feedback policies for robust control with constraints. *Automatica*, 42(4):523–533, 2006.

G. Grimm, M. J. Messina, S. E. Tuna, and A. R. Teel. Examples when nonlinear model predictive control is nonrobust. *Automatica*, 40(10):1729–1738, 2004.

G. Grimm, M. J. Messina, S. E. Tuna, and A. R. Teel. Nominally robust model predictive control with state constraints. *IEEE Transactions on Automatic Control*, 52(10):1856–1870, 2007.

S. Gros. An economic NMPC formulation for wind turbine control. In *Proceedings of the 52nd IEEE Conference on Decision and Control*, pages 1001–1006, 2013.

J. M. Grosso, C. Ocampo-Martinez, and V. Puig. Reliability–based economic model predictive control for generalised flow–based networks including actuators' health–aware capabilities. *International Journal of Applied Mathematics and Computer Science*, 26(3): 641–654, 2016.

L. Grüne. Economic receding horizon control without terminal constraints. *Automatica*, 49 (3):725–734, 2013.

L. Grüne and M. A. Müller. On the relation between strict dissipativity and turnpike properties. *Systems & Control Letters*, 90:45–53, 2016.

L. Grüne and J. Pannek. *Nonlinear Model Predictive Control*. Springer, 2011.

L. Grüne and M. Stieler. Asymptotic stability and transient optimality of economic MPC without terminal conditions. *Journal of Process Control*, 24(8):1187–1196, 2014.

L. Grüne and M. Zanon. Periodic optimal control, dissipativity and MPC. In *Proceedings of the 21st international symposium on mathematical theory of networks and systems*, pages 1804–1807, 2014.

M. Guay, V. Adetola, and D. DeHaan. *Robust and Adaptive Model Predictive Control of Nonlinear Systems*. Institution of Engineering and Technology, 2015.

F. Guerra Vázquez, J.-J. Rückmann, O. Stein, and G. Still. Generalized semi-infinite programming: A tutorial. *Journal of Computational and Applied Mathematics*, 217(2): 394–419, 2008.

R. Halvgaard, N. K. Poulsen, H. Madsen, J. B. Jørgensen, F. Marra, and D. E. M. Bondy. Electric vehicle charge planning using economic model predictive control. In *Proceedings of the IEEE International Electric Vehicle Conference*, pages 1–6, 2012.

M. A. Henson and D. E. Seborg, editors. *Nonlinear Process Control*. Prentice-Hall, Inc., 1997.

D. H. V. Hessem, C. W. Scherer, and O. H. Bosgra. LMI-based closed-loop economic optimization of stochastic process operation under state and input constraints. In *Proceedings of the 40th IEEE Conference on Decision and Control*, pages 4228–4233, 2001.

B. Houska. *Robust Optimization of Dynamic Systems*. PhD thesis, Faculty of Engineering, Katholieke Universiteit Leuven, Belgium, 2011.

T. G. Hovgaard, L. F. S. Larsen, and J. B. Jorgensen. Robust economic MPC for a power management scenario with uncertainties. In *Proceedings of the 50th IEEE Conference on Decision and Control and European Control Conference*, pages 1515–1520, 2011.

R. Huang, L. T. Biegler, and E. Harinath. Robust stability of economically oriented infinite horizon NMPC that include cyclic processes. *Journal of Process Control*, 22(1):51–59, 2012.

M. Z. Jamaludin and C. L. Swartz. A bilevel programming formulation for dynamic real-time optimization. In *Proceedings of the 9th IFAC Symposium on Advanced Control of Chemical Processes*, pages 906–911, 2015.

Z.-P. Jiang and Y. Wang. Input-to-state stability for discrete-time nonlinear systems. *Automatica*, 37(6):857–869, 2001.

Z.-P. Jiang and Y. Wang. A converse Lyapunov theorem for discrete-time systems with disturbances. *Systems & Control Letters*, 45(1):49–58, 2002.

J. B. Jørgensen, L. E. Sokoler, L. Standardi, R. Halvgaard, T. G. Hovgaard, G. Frison, N. K. Poulsen, and H. Madsen. Economic MPC for a linear stochastic system of energy units. In *Proceedings of the European Control Conference*, pages 903–909, 2016.

J. V. Kadam and W. Marquardt. Integration of economical optimization and control for intentionally transient process operation. In R. Findeisen, F. Allgöwer, and L. T. Biegler, editors, *Assessment and Future Directions of Nonlinear Model Predictive Control*, volume 358 of *Lecture Notes in Control and Information Sciences*, pages 419–434. Springer Berlin Heidelberg, 2007.

C. M. Kellett. A compendium of comparison function results. *Mathematics of Control, Signals, and Systems*, 26(3):339–374, 2014.

E. C. Kerrigan. *Robust Constraint Satisfaction: Invariant Sets and Predictive Control*. PhD thesis, Department of Engineering, University of Cambridge, UK, 2000.

E. C. Kerrigan and J. M. Maciejowski. Feedback min-max model predictive control using a single linear program: Robust stability and the explicit solution. *International Journal of Robust and Nonlinear Control*, 14(4):395–413, 2004.

H. K. Khalil. *Nonlinear Systems*. Prentice-Hall, Inc., third edition, 2002.

A. Klenke. *Probability Theory: A Comprehensive Course*. Universitext. Springer, second edition, 2014.

I. Kolmanovsky and E. G. Gilbert. Maximal output admissible sets for discrete-time systems with disturbance inputs. In *Proceedings of the American Control Conference*, volume 3, pages 1995–1999, 1995.

I. Kolmanovsky and E. G. Gilbert. Theory and computation of disturbance invariant sets for discrete-time linear systems. *Mathematical Problems in Engineering*, 4(4):317–367, 1998.

M. Korda, R. Gondhalekar, J. Cigler, and F. Oldewurtel. Strongly feasible stochastic model predictive control. In *Proceedings of the 50th IEEE Conference on Decision and Control and European Control Conference*, pages 1245–1251, 2011.

K. I. Kouramas, S. V. Raković, E. C. Kerrigan, J. C. Allwright, and D. Q. Mayne. On the minimal robust positively invariant set for linear difference inclusions. In *Proceedings of the 44th IEEE Conference on Decision and Control*, pages 2296–2301, 2005.

B. Kouvaritakis and M. Cannon. *Model Predictive Control: Classical, Robust and Stochastic.* Springer, 2015.

B. Kouvaritakis, M. Cannon, S. V. Raković, and Q. Cheng. Explicit use of probabilistic distributions in linear predictive control. *Automatica*, 46(10):1719–1724, 2010.

M. Kvasnica, P. Grieder, and M. Baotić. Multi-Parametric Toolbox 2.6.2 (MPT), 2004. URL http://control.ee.ethz.ch/~mpt/2/.

W. Langson, I. Chryssochoos, S. V. Raković, and D. Q. Mayne. Robust model predictive control using tubes. *Automatica*, 40(1):125–133, 2004.

M. Lazar, D. Muñoz de la Peña, W. P. M. H. Heemels, and T. Alamo. On input-to-state stability of min-max nonlinear model predictive control. *Systems & Control Letters*, 57 (1):39–48, 2008.

J. Lee and Z. Yu. Worst-case formulations of model predictive control for systems with bounded parameters. *Automatica*, 33(5):763–781, 1997.

D. Limón, T. Alamo, and E. Camacho. Input-to-state stable MPC for constrained discrete-time nonlinear systems with bounded additive uncertainties. In *Proceedings of the 41st IEEE Conference on Decision and Control*, pages 4619–4624, 2002a.

D. Limón, T. Alamo, and E. F. Camacho. Stability analysis of systems with bounded additive uncertainties based on invariant sets: Stability and feasibility of MPC. In *Proceedings of the American Control Conference*, pages 364–369, 2002b.

D. Limón, J. Bravo, T. Alamo, and E. Camacho. Robust MPC of constrained nonlinear systems based on interval arithmetic. *IEE Proceedings - Control Theory and Applications*, 152(3):325–332, 2005.

D. Limón, T. Alamo, F. Sala, and E. Camacho. Input to state stability of min-max MPC controllers for nonlinear systems with bounded uncertainties. *Automatica*, 42:797–803, 2006.

D. Limón, T. Alamo, D. Raimondo, D. Muñoz de la Peña, J. Bravo, A. Ferramosca, and E. Camacho. Input-to-state stability: A unifying framework for robust model predictive control. In L. Magni, D. Raimondo, and F. Allgöwer, editors, *Nonlinear Model Predictive Control*, volume 384 of *Lecture Notes in Control and Information Sciences*, pages 1–26. Springer Berlin Heidelberg, 2009.

D. Limón, M. Pereira, D. Muñoz de la Peña, and J. G. T. Alamo. Single-layer economic model predictive control for periodic operation. *Journal of Process Control*, 24(8): 1207–1224, 2014.

S. Liu, J. Zhang, and J. Liu. Economic MPC with terminal cost and application to an oilsand primary separation vessel. *Chemical Engineering Science*, 136:27–37, 2015.

J. Löfberg. Approximations of closed-loop minimax MPC. In *Proceedings of the 52nd IEEE Conference on Decision and Control*, pages 1438–1442, 2003.

M. Lorenzen, F. Allgöwer, F. Dabbene, and R. Tempo. An improved constraint-tightening approach for stochastic MPC. In *Proceedings of the American Control Conference*, pages 944–949, 2015.

M. Lorenzen, F. Allgöwer, and M. Cannon. Adaptive model predictive control with robust constraint satisfaction. In *Proceedings of the 20th IFAC World Congress*, pages 3368–3373, 2017a.

M. Lorenzen, M. A. Müller, and F. Allgöwer. Stochastic model predictive control without terminal constraints. *International Journal of Robust and Nonlinear Control*, 2017b. doi: 10.1002/rnc.3912. to appear.

S. Lucia and S. Engell. Control of towing kites under uncertainty using robust economic nonlinear model predictive control. In *Proceedings of the European Control Conference*, pages 1158–1163, 2014.

S. Lucia, J. A. Andersson, H. Brandt, M. Diehl, and S. Engell. Handling uncertainty in economic nonlinear model predictive control: A comparative case study. *Journal of Process Control*, 24(8):1247–1259, 2014.

W.-J. Ma and V. Gupta. Desynchronization of thermally-coupled first-order systems using economic model predictive control. In *Proceedings of the 51st IEEE Conference on Decision and Control*, pages 278–283, 2012.

Y. Ma, A. Kelman, A. Daly, and F. Borrelli. Predictive control for energy efficient buildings with thermal storage: Modeling, simulation, and experiments. *IEEE Control Systems*, 32 (1):44–64, 2012.

L. Magni and R. Scattolini. Robustness and robust design of MPC for nonlinear discrete-time systems. In R. Findeisen, F. Allgöwer, and L. T. Biegler, editors, *Assessment and Future Directions of Nonlinear Model Predictive Control*, pages 239–254. Springer Berlin Heidelberg, Berlin, Heidelberg, 2007.

L. Magni, D. M. Raimondo, and R. Scattolini. Regional input-to-state stability for nonlinear model predictive control. *IEEE Transactions on Automatic Control*, 51(9):1548–1553, 2006.

L. Magni, D. M. Raimondo, and F. Allgöwer, editors. *Nonlinear Model Predictive Control: Towards New Challenging Applications*. Springer, 2009.

W. Mai and C. Chung. Economic MPC of aggregating commercial buildings for providing flexible power reserve. *IEEE Transactions on Power Systems*, 30(5):2685–2694, 2015.

J. P. Maree and L. Imsland. Combined economic and regulatory predictive control. *Automatica*, 69:342–347, 2016.

A. Marquez, J. Patiño, and J. Espinosa. Min-max economic model predictive control. In *Proceedings of the 53rd IEEE Conference on Decision and Control*, pages 4410–4415, 2014.

MATLAB. *Optimization Toolbox Release 2015b*. The MathWorks, Inc., Natick, Massachusetts, 2015.

D. Mayne, S. Raković, R. Findeisen, and F. Allgöwer. Robust output feedback model predictive control of constrained linear systems. *Automatica*, 42(7):1217–1222, 2006.

D. Q. Mayne. Control of constrained dynamic systems. *European Journal of Control*, 7(2): 87–99, 2001.

D. Q. Mayne. Model predictive control: Recent developments and future promise. *Automatica*, 50(12):2967–2986, 2014.

D. Q. Mayne. Robust and stochastic model predictive control: Are we going in the right direction? *Annual Reviews in Control*, 41:184–192, 2016.

D. Q. Mayne, J. B. Rawlings, C. V. Rao, and P. O. M. Scokaert. Constrained model predictive control: Stability and optimality. *Automatica*, 36(6):789–814, 2000.

D. Q. Mayne, M. M. Seron, and S. V. Raković. Robust model predictive control of constrained linear systems with bounded disturbances. *Automatica*, 41(2):219–224, 2005.

D. Q. Mayne, E. C. Kerrigan, E. J. van Wyk, and P. Falugi. Tube-based robust nonlinear model predictive control. *International Journal of Robust and Nonlinear Control*, 21: 1341–1353, 2011.

A. Mesbah. Stochastic model predictive control: An overview and perspectives for future research. *IEEE Control Systems*, 36(6):30–44, 2016.

A. Mesbah, S. Streif, R. Findeisen, and R. D. Braatz. Stochastic nonlinear model predictive control with probabilistic constraints. In *Proceedings of the American Control Conference*, pages 2413–2419, 2014.

H. Michalska and D. Q. Mayne. Robust receding horizon control of constrained nonlinear systems. *IEEE Transactions on Automatic Control*, 38(11):1623–1633, 1993.

L. M. Miller, Y. Silverman, M. A. MacIver, and T. D. Murphey. Ergodic exploration of distributed information. *IEEE Transactions on Robotics*, 32(1):36–52, 2016.

D. Muñoz de la Peña, A. Bemporad, and T. Alamo. Stochastic programming applied to model predictive control. In *Proceedings of the 44th IEEE Conference on Decision and Control and European Control Conference*, pages 1361–1366, 2005.

M. A. Müller. *Distributed and economic model predictive control: beyond setpoint stabilization*. PhD thesis, Faculty of Engineering Design, Production Engineering and Automotive Engineering, University of Stuttgart, Germany, 2014.

M. A. Müller and F. Allgöwer. Robustness of steady-state optimality in economic model predictive control. In *Proceedings of the 41st IEEE Conference on Decision and Control*, pages 1011–1016, 2012.

M. A. Müller and F. Allgöwer. Economic and distributed model predictive control: Recent developments in optimization-based control. *SICE Journal of Control, Measurement, and System Integration*, 10(2):39–52, 2017.

M. A. Müller and L. Grüne. Economic model predictive control without terminal constraints for optimal periodic behavior. *Automatica*, 70:128–139, 2016.

M. A. Müller, D. Angeli, and F. Allgöwer. On convergence of averagely constrained economic MPC and necessity of dissipativity for optimal steady-state operation. In *Proceedings of the American Control Conference*, pages 3141–3146, 2013.

M. A. Müller, D. Angeli, and F. Allgöwer. Transient average constraints in economic model predictive control. *Automatica*, 50(11):2943–2950, 2014a.

M. A. Müller, D. Angeli, F. Allgöwer, R. Amrit, and J. B. Rawlings. Convergence in economic model predictive control with average constraints. *Automatica*, 50(12):3100–3111, 2014b.

M. A. Müller, D. Angeli, and F. Allgöwer. On necessity and robustness of dissipativity in economic model predictive control. *IEEE Transactions on Automatic Control*, 60(6):1671–1676, 2015a.

M. A. Müller, L. Grüne, and F. Allgöwer. On the role of dissipativity in economic model predictive control. In *Proceedings of the 5th IFAC Conference on Nonlinear Model Predictive Control*, pages 110–116, 2015b.

J. R. Munkres. *Topology*. Featured Titles for Topology Series. Prentice-Hall, Inc., second edition, 2000.

D. Muñoz-Carpintero, B. Kouvaritakis, and M. Cannon. Striped parameterized tube model predictive control. *Automatica*, 67:303–309, 2016.

C.-J. Ong and E. G. Gilbert. The minimal disturbance invariant set: Outer approximations via its partial sums. *Automatica*, 42(9):1563–1568, 2006.

M. Ono and B. C. Williams. Iterative risk allocation: A new approach to robust model predictive control with a joint chance constraint. In *Proceedings of the 47th IEEE Conference on Decision and Control*, pages 3427–3432, 2008.

B. K. Pagnoncelli, S. Ahmed, and A. Shapiro. Sample average approximation method for chance constrained programming: Theory and applications. *Journal of Optimization Theory and Applications*, 142(2):399–416, 2009.

Q. Pang, T. Zou, Q. Cong, and Y. Wang. Constrained model predictive control with economic optimization for integrating process. *The Canadian Journal of Chemical Engineering*, 93(8):1462–1473, 2015.

G. Pannocchia and E. C. Kerrigan. Offset-free receding horizon control of constrained linear systems. *AIChE Journal*, 51(12):3134–3146, 2005.

G. Pannocchia, J. B. Rawlings, and S. J. Wright. Inherently robust suboptimal nonlinear MPC: Theory and application. In *Proceedings of the 50th IEEE Conference on Decision and Control and European Control Conference*, pages 3398–3403, 2011a.

G. Pannocchia, J. B. Rawlings, and S. J. Wright. Conditions under which suboptimal nonlinear MPC is inherently robust. *Systems & Control Letters*, 60(9):747–755, 2011b.

N. R. Patel, M. J. Risbeck, J. B. Rawlings, M. J. Wenzel, and R. D. Turney. Distributed economic model predictive control for large-scale building temperature regulation. In *Proceedings of the American Control Conference*, pages 895–900, 2016.

M. Pereira, D. Muñoz de la Peña, D. Limón, I. Alvarado, and T. Alamo. Robust model predictive controller for tracking changing periodic signals. *IEEE Transactions on Automatic Control*, 62(10):5343–5350, 2016.

G. Pin, D. M. Raimondo, L. Magni, and T. Parisini. Robust model predictive control of nonlinear systems with bounded and state-dependent uncertainties. *IEEE Transactions on Automatic Control*, 54(7):1681–1687, 2009.

A. S. Poznyak. *Advanced Mathematical Tools for Automatic Control Engineers, 2: Stochastic Systems*. Elsevier, 2009.

J. A. Primbs and C. H. Sung. Stochastic receding horizon control of constrained linear systems with state and control multiplicative noise. *IEEE Transactions on Automatic Control*, 54(2):221–230, 2009.

V. Puig, R. Costa-Castelló, and J. L. Sampietro. Economic MPC for the energy management of hybrid vehicles including fuel cells and supercapacitors. In *Proceedings of the 11th UKACC International Conference on Control*, pages 1–6, 2016.

D. E. Quevedo, G. C. Goodwin, and J. A. De Doná. Finite constraint set receding horizon quadratic control. *International Journal of Robust and Nonlinear Control*, 14(4):355–377, 2004.

D. M. Raimondo, T. Alamo, D. Limón, and E. F. Camacho. Towards the practical implementation of min-max nonlinear model predictive control. In *Proceedings of the 46th IEEE Conference on Decision and Control*, pages 1257–1262, 2007.

D. M. Raimondo, D. Limón, M. Lazar, L. Magni, and E. F. Camacho. Min-max model predictive control of nonlinear systems: A unifying overview on stability. *European Journal of Control*, 15(1):5–21, 2009.

S. V. Raković and M. Barić. Parameterized robust control invariant sets for linear systems: Theoretical advances and computational remarks. *IEEE Transactions on Automatic Control*, 55(7):1599–1614, 2010.

S. V. Raković, E. C. Kerrigan, K. I. Kouramas, and D. Q. Mayne. Invariant approximations of the minimal robust positively invariant set. *IEEE Transactions on Automatic Control*, 50(3):406–410, 2005.

S. V. Raković, A. R. Teel, D. Q. Mayne, and A. Astolfi. Simple robust control invariant tubes for some classes of nonlinear discrete time systems. In *Proceedings of the 45th IEEE Conference on Decision and Control*, pages 6397–6402, 2006.

S. V. Raković, E. C. Kerrigan, D. Q. Mayne, and K. I. Kouramas. Optimized robust control invariance for linear discrete-time systems: Theoretical foundations. *Automatica*, 43(5): 831–841, 2007.

S. V. Raković, B. Kouvaritakis, M. Cannon, and C. Panos. Fully parameterized tube model predictive control. *International Journal of Robust and Nonlinear Control*, 22(12): 1330–1361, 2012a.

S. V. Raković, B. Kouvaritakis, M. Cannon, C. Panos, and R. Findeisen. Parameterized tube model predictive control. *IEEE Transactions on Automatic Control*, 57(11):2746–2761, 2012b.

S. V. Raković, B. Kouvaritakis, R. Findeisen, and M. Cannon. Homothetic tube model predictive control. *Automatica*, 48(8):1631–1638, 2012c.

S. V. Raković, B. Kouvaritakis, and M. Cannon. Equi-normalization and exact scaling dynamics in homothetic tube model predictive control. *Systems & Control Letters*, 62(2): 209–217, 2013.

D. R. Ramirez and E. F. Camacho. On the piecewise linear nature of constrained min-max model predictive control with bounded uncertainties. In *Proceedings of the American Control Conference*, pages 3620–3625, 2003.

J. B. Rawlings and R. Amrit. Optimizing process economic performance using model predictive control. In L. Magni, D. M. Raimondo, and F. Allgöwer, editors, *Nonlinear Model Predictive Control: Towards New Challenging Applications*, volume 384 of *Lecture Notes in Control and Information Sciences*, pages 119–138. Springer Berlin Heidelberg, 2009.

J. B. Rawlings and D. Q. Mayne. *Model Predictive Control: Theory and Design*, volume 1. Nob Hill, 2009.

J. B. Rawlings and D. Q. Mayne. Postface to Model Predictive Control: Theory and Design, 2012. Available online: http://jbrwww.che.wisc.edu/home/jbraw/mpc/postface.pdf.

J. B. Rawlings and M. J. Risbeck. Model predictive control with discrete actuators: Theory and application. *Automatica*, 78:258–265, 2017.

J. B. Rawlings, D. Angeli, and C. N. Bates. Fundamentals of economic model predictive control. In *Proceedings of the 51st IEEE Conference on Decision and Control*, pages 3851–3861, 2012.

M. Reble and F. Allgöwer. Unconstrained model predictive control and suboptimality estimates for nonlinear continuous-time systems. *Automatica*, 48(8):1812–1817, 2012.

M. Reble, D. E. Quevedo, and F. Allgöwer. Improved stability conditions for unconstrained nonlinear model predictive control by using additional weighting terms. In *Proceedings of the 51st IEEE Conference on Decision and Control*, pages 2625–2630, 2012.

S. Revollar, P. Vega, M. Francisco, R. Vilanova, and I. Santín. Optimization of economic and environmental objectives in a non linear model predictive control applied to a wastewater treatment plant. In *Proceedings of the 20th International Conference on System Theory, Control and Computing*, pages 318–323, 2016.

R. T. Rockafellar and R. J.-B. Wets. *Variational analysis*, volume 317. Springer Science & Business Media, 2009.

A. T. Schwarm and M. Nikolaou. Chance-constrained model predictive control. *AIChE Journal*, 45(8):1743–1752, 1999.

P. O. M. Scokaert and D. Q. Mayne. Min-max feedback model predictive control for constrained linear systems. *IEEE Transactions on Automatic Control*, 43(8):1136–1142, 1998.

L. Seban, N. Boruah, and B. Roy. Modified single layer economic model predictive control and application to shell and tube heat exchanger. In *Proceedings of the 4th IFAC Conference on Advances in Control and Optimization of Dynamical Systems*, pages 754–759, 2016.

S. E. Shafiei, J. Stoustrup, and H. Rasmussen. A supervisory control approach in economic MPC design for refrigeration systems. In *Proceedings of the European Control Conference*, pages 1565–1570, 2013.

M. L. Shaltout, Z. Ma, and D. Chen. An economic model predictive control approach using convex optimization for wind turbines. In *Proceedings of the American Control Conference*, pages 3176–3181, 2016.

L. E. Sokoler, K. Edlund, and J. B. Jørgensen. Application of economic MPC to frequency control in a single-area power system. In *Proceedings of the 54th IEEE Conference on Decision and Control*, pages 2635–2642, 2015.

E. D. Sontag. Smooth stabilization implies coprime factorization. *IEEE Transactions on Automatic Control*, 34(4):435–443, 1989.

E. D. Sontag and Y. Wang. On characterizations of the input-to-state stability property. *Systems & Control Letters*, 24(5):351–359, 1995.

P. Sopasakis, D. Herceg, P. Patrinos, and A. Bemporad. Stochastic economic model predictive control for Markovian switching systems. In *Proceedings of the 20th IFAC World Congress*, pages 526–532, 2017.

A. Staino, H. Nagpal, and B. Basu. Cooperative optimization of building energy systems in an economic model predictive control framework. *Energy and Buildings*, 128:713–722, 2016.

K. Subramanian, J. B. Rawlings, and C. T. Maravelias. Economic model predictive for inventory management in supply chains. *Computers & Chemical Engineer* 71–80, 2014.

M. Tanaskovic, L. Fagiano, R. Smith, and M. Morari. Adaptive receding horizon co for constrained MIMO systems. *Automatica*, 50(12):3019–3029, 2014.

F. Tedesco, L. Mariam, M. Basu, A. Casavola, and M. F. Conlon. Supervision of commun based microgrids: An economic model predictive control approach. In *Proceedings c the International Conference on Renewable Energies and Power Quality*, pages 172–177 2016.

A. R. Teel. Discrete time receding horizon optimal control: Is the stability robust? In M. S. de Queiroz, M. Malisoff, and P. Wolenski, editors, *Optimal Control, Stabilization and Nonsmooth Analysis*, volume 301 of *Lecture Notes in Control and Information Sciences*, pages 3–27. Springer Berlin Heidelberg, 2004.

H. Tian, Q. Lu, R. B. Gopaluni, V. M. Zavala, and J. A. Olson. Economic nonlinear model predictive control for mechanical pulping processes. In *Proceedings of the American Control Conference*, pages 1796–1801, 2016.

K. P. Wabersich. Robust economic model predictive control for periodic operation with application to supply chain networks. Master's thesis, Institute for Systems Theory and Automatic Control, University of Stuttgart, 2017.

Y. Wang and S. Boyd. Performance bounds for linear stochastic control. *Systems & Control Letters*, 58(3):178–182, 2009.

Y. Wang, T. Alamo, V. Puig, and G. Cembrano. Periodic economic model predictive control with nonlinear-constraint relaxation for water distribution networks. In *Proceedings of the IEEE Conference on Control Applications*, pages 1167–1172, 2016.

J. C. Willems. Dissipative dynamical systems part I: General theory. *Archive for Rational Mechanics and Analysis*, 45(5):321–351, 1972.

Z. Wu, M. Cui, X. Xie, and P. Shi. Theory of stochastic dissipative systems. *IEEE Transactions on Automatic Control*, 56(7):1650–1655, 2011.

L. Würth, R. Hannemann, and W. Marquardt. A two-layer architecture for economically optimal process control and operation. *Journal of Process Control*, 21(3):311–321, 2011.

S. Yu, M. Reble, H. Chen, and F. Allgöwer. Inherent robustness properties of quasi-infinite horizon MPC. In *Proceedings of the 18th IFAC World Congress*, pages 179–184, 2011.

S. Yu, C. Maier, H. Chen, and F. Allgöwer. Tube MPC scheme based on robust control invariant set with application to Lipschitz nonlinear systems. *Systems & Control Letters*, 62(2):194–200, 2013.

S. Yu, M. Reble, H. Chen, and F. Allgöwer. Inherent robustness properties of quasi-infinite horizon nonlinear model predictive control. *Automatica*, 50(9):2269–2280, 2014.

liography

control
9, 64:

. Nonlinear economic model predictive control for microgrid
s of the 10th IFAC Symposium on Nonlinear Control Systems,

trol

and M. Diehl. A Lyapunov function for periodic economic optimizing
control. In *Proceedings of the 52nd IEEE Conference on Decision and*
5107–5112, 2013.

. Gros, and M. Diehl. A tracking MPC formulation that is locally equivalent
mic MPC. *Journal of Process Control*, 45:30–42, 2016a.

.on, L. Grüne, and M. Diehl. Periodic optimal control, dissipativity and MPC. *IEEE*
ansactions on Automatic Control, 62(6):2943–2949, 2016b.

. M. Zavala. A multiobjective optimization perspective on the stability of economic MPC.
In *Proceedings of the 9th IFAC Symposium on Advanced Control of Chemical Processes*,
pages 974–980, 2015.

J. Zeng and J. Liu. Economic model predictive control of wastewater treatment processes.
Industrial & Engineering Chemistry Research, 54(21):5710–5721, 2015.

X. Zhang, S. Grammatico, G. Schildbach, P. Goulart, and J. Lygeros. On the sample size
of randomized MPC for chance-constrained systems with application to building climate
control. In *Proceedings of the European Control Conference*, pages 478–483, 2014.

Z. Q. Zheng and M. Morari. Robust stability of constrained model predictive control. In
Proceedings of the American Control Conference, pages 379–383, 1993.

Q. Zhu, S. Onori, and R. Prucka. Nonlinear economic model predictive control for SI
engines based on sequential quadratic programming. In *Proceedings of the American*
Control Conference, pages 1802–1807, 2016.

Y. Zong, G. M. Böning, R. M. Santos, S. You, J. Hu, and X. Han. Challenges of implementing
economic model predictive control strategy for buildings interacting with smart energy
systems. *Applied Thermal Engineering*, 114:1476–1486, 2017.

V. A. Zorich. *Mathematical Analysis II*, volume 65. Springer, 2004.